PHYSIOLOGY OF TRANSPORT IN PLANTS

PHYSIOLOGY OF TRANSPORT IN PLANTS

By

DR. S. SUNDARA RAJAN
Principal
Reva Institute for Science and Technology Studies
R.T. Nagar
Bangalore – 560 024

ANMOL PUBLICATIONS PVT. LTD.
NEW DELHI - 110 002 (INDIA)

ANMOL PUBLICATIONS PVT. LTD.
4374/4B, Ansari Road, Daryaganj
New Delhi - 110 002
Ph.: 3261597, 3278000
Visit us at: www.anmolbooks.com

Physiology of Transport in Plants

First Edition, 2003

ISBN 81-261-1362-6

PRINTED IN INDIA

Published by J.L. Kumar for Anmol Publications Pvt. Ltd., New Delhi - 110 002 and Printed at Tarun Offset.

CONTENTS

Preface

Plants are unique among living beings as they are capable of producing enough food materials not only for themsleves but for other organisms also. They remain *in situ* ,absorb the necessary raw materials from soil and manufacture organic substances . The stationary posture of plants however poses a problem for plants as all parts can not absorb inorganic raw materials from soil and also underground parts (being non green) can not prepare food . This necessiates a mechanism for multichannnel transportation. The phenomenon of transportation in deed becomes very complex when distances to be travelled are very long as in trees.

'PHYSIOLOGY OF TRANSPORT IN PLANTS' analyses the various aspects of transportation such as structures involved , direction , pathways, mechanism , energy involved etc in a detailed way.

The book has chapters dealing with transnportation phenomena of both organic and inorganic constituents including the absorption and loss of water .As the understanding of cell structure is intrergral to the study of any physiological process, chapter two deals with cell in an exhaustive manner.

The book is intended for both gradute and post graduate students taking Biology, Botnay as a core subject.

I am thankful to Mr J.L Kumar, M.D. of Anmol Publications, New Delhi for publishing this book.

Dr. S. SUNDARA RAJAN
Principal

Reva Institute For Science and Technology studies
R.T. Nagar
Bangalore - 560 024

1

INTRODUCTION

Green plants in general and higher vascular plants in particular provide sustenance for all forms of life including human. Plants play a vital role in the survival of man on land. Human life as we known it perhaps would have been impossible but for plants. Plants provide us the three basic necessities of life namely food, clothing and shelter. In order to benefit from plant life, a thorough knowledge of their growth, development and function is necessary. For, it is this knowledge that can help us tomake attempts towards improvement of plant life. The branch of botany that deals with various functional aspects of plants is called **plant physiology.**

Plants physiology may be said to be the study of vital phenomena in plants. It is concerned with the processes and functions, and also responses of plants to their surroundings. The result of plant's responses to the environment namely growth, development and reproduction are also studied in plant physiology.

The processes studied in plant physiology are photosynthesis, respiration, translocation etc. Each process has a continuing sequence of events and plant physiology explains and describes each one of these events. Plant physiology also attempts to explain the functions of plants at various levels. Functions of cell organelles, cells, tissues and organs are explained.

To summarize, the following are the main tasks of plant physiology.

1. Describe and explain processes.
2. Describe and explain the functions at various levels in the plant organization.
3. Explain the function of chemical constituents present in the body of plants.
4. Explain the effects of the environment on processes and functions in the plant system.
5. Emphasize the interrelationship and interdependence of various processes and functions, so as to understand the life processes going on in the plant.

In order to realise the above objectives, plant physiologists derive help from physics, chemistry, mathematics, statistics etc. Indeed, the role of these physical sciences is greater in plant physiology than in other branch of biology. Modern developments in these branches of physical sciences have greatly helped in unravelling the secrets of many an important physiological process in plants.

Various tools and techniques are employed, both physical and biological, in the study of plant physiology. Microscope, Chromatography, Autoradiography, Electrophoresis, Tracer techniques, Spectrophotometry and a whole range of biochemical techniques form a part and parcel of the instrumental arsenal of a plant physiologist. Plant physiology is a dynamic discipline which continuously interacts with other disciplines and constantly updates itself in tune with advancements in other disciplines of sciences. Developments in the field of biotechnology, genetic engineering etc., have also helped the study of various physiological processes *in vitro*.

Many of the physiological processes are common between plants and animals at the cellular and molecular level. Energy transformation reactions, transport of substances across the membrane are some activities common between plants and animals. But plants physiology differs from animal physiology in many of the vital processes. Plants and animals have two basic differences in their mode of life - locomotion and method of obtaining nutrition. These two fundamental facts are primarily responsible for thedifferences in physiology between plants and animals. Animals are like machines not only in function, but even in their morphology. They have a genetic rigidty in size and shape, while plants grow and operate on an architectural basis, which is beautifully ambiguous to allow sufficient plasticity on a basic infrastructure. The following is a list of unique characters of plants responsible for their special physiological pattern. (The list is based on - Plant physiology; R.G.S. Bidwell, 1979).

1. Plants are non motile as such they can exploit only a limited volume of their environment.
2. Carbon autotrophy permits a variety of carbon metabolism.
3. As plants are dependent on soil for nutrition, they are highly conservative with reference to mineral nutrition, specially so in concerning with nitrogen.
4. As plants have evolved to a terrestrial habit at various mechanisms have been evolved that protect them from water loss and permit them to carry on gaseous exchange without actual breathing.
5. Efficient transportation mechanism has been evolved mainly due to adaptation for terrestrial habit.

6. The structural support in plants mainly provided by the cell wall (in lieu of skeletal system found in animals) has posed many problems.
7. There are more number of interconnections between plants cells than among animal cells.
8. Plants face serious seasonal problems. They have accurate time measuring devices to co-ordinate their activities to the seasonal changes of the environment. Germination, dormacy, leaf fall, flowering etc., are all seasonal.
9. As plants are immobile, they cannot seek protection from adversaries by escaping. Hence they have special means of protection from adverse elements of nature like wind, cold, heat, drought etc.
10. Since plants have an indeterminate growth, they are endowed with highly complex physiological and biochemical mechanisms.
11. There is no organised nervous system in plants but they have a biochemical means of communication between their parts. This results in further elaboration of biochemical patern. (It is however not certain even now, as to how and in what form message impulses are transported within the plants. It is also not known whether there is any communication at all between plants).

Practical aspects of plant physiology

Why study plant physiology? What are the benefits that accrue from a study of plant physiology? These are some of the questions that we will attempt to anser here.

The study of plant physiology is necessary not only from the theoretical stand point in helpingo us to understand the basic functioning of the plant, but also from the point of view of application of knowledge of plant physiology for human welfare. Agronomy, Horticulture, Forestry, floriculture, plant breeding, plant pathology, pharmacology are some of the applied fields of science, where a basic understanding of plant physiology is necessary.

Modern agricultural practices are no longer the concern of only the farmer. It is now in the realm of intense research in laboratories. Mankind is faced with the problem of feeding the ever increasing population with a finite resource. Hence it is imperative to increase the yield in the available space. In helping to understand soil fertility, growth of plants and in improving the efficiency of plants, the knowledge of plant physiology is a basic necessity. Mention of once or two instances where plant physiology directly affects food production perhaps may underscore the points mentioned above.

(a) Research in photosynthesis, can increase photosynthetic efficiency by reducing photorespiration. Breeding program may be planned to alter the leaf exposure to sunlight resulting in increased light absorption.

(b) Improvement of biological N_2 fixation will result in reduced input of costly chemical fertilizers.

(c) Techniques of tissues culture and protoplast fusion have resulted in the production of disease free and new varieties of plants.

(d) Frost resisting, drought resisting and disease resistant varieties have been developed.

(e) Research in water stress and antitranspirants have resulted in maximum water economy in plants.

Relationship of plant physiology with other sciences

Plant physiology mainly deals with the functions of plant organs and the physical and chemical changes associated with them. These can be explained on the basis of known physical and chemical principles. Hence a thorough knowledge of the fundamental principles of physics and chemistry is very much necessary for the understanding of plant physiology. Biochemistry is another science together with Biophysics, whose knowledge is necessary for plant physiology. The knowledge of biochemistry helps us to isolate, purify and understand the metabolic pathways of various chemical constituents of plants, while biophysics helps us to understand molecular structure of the various components occurring in plants.

Within biology, Anatomy, Genetics and Ecology (physiological plant anatomy, physiological genetics and physiological ecology) are three of the branches, whose interest and subject matter overlaps that of plant physiology. In a way physiology is the fulcrum on which revolve all other branches of biology.

2

THE CELL

One of the marvels of nature is the creation of tiny yet potent biological compartments called cells. There is virtually no living being in the biological world save for Viruses which is not cellular. Either one, two or many millions of cells go to make the body of the living beings. A cell may be defined as the 'structural and functional unit of a living being. It is the minimal biological unit capable of maintaining and propagating itself. The study of cell comes under the perview of cytology, physiology and anatomy; while anatomy deals mainly with the structural aspects, the other two lay emphasis both on structure and function.

History

The word cell has been derived from the Latin word *cellula* meaning a small compartment. The term in the present sense was first used by Robert Hooke (1665) in his book *Micrographic.*

There has been some mention about cell in ancient Indian literature also. Parasara (1st century BC) in his book *Vrkshayurveda* refers to invisible compartments in the leaf and names them *Rasakc ha.* He further opines that each cell has a sap *(Rasa),* not visible to the naked eye *(Anavah)* and has two boundaries *(Kalavestitena),* some of the cells have a colouring matter *(Ranjakayukta).*

The credit however of discovering the cell goes to Robert Hooke (1965), since it s difficult to establish the discovery to the ancient Indians in the absence of complete text of Parasara and the general apathy among Indians to their scientific heritage. Robert Hooke who constructed the first compound microscope observed the sections of cork and opined that they contain honey comb like compartments. German Biologists M.J. Schleiden and T.S. Schwann(1838) established the '*cell theory*' that all organisms are made up of cells.

Dutrochet in 1824 separated the cells from *Mimosa* plant. He boiled the leaves in nitric acid to isolate the cells. Turpin and Meyers (1830) observed that every tissue is an assemblage of cells.

One of the significant discoveries of the cell came from Robert Brown (1830). He discovered the presence of a spherical body in the centre of every cell, which he named *'nucleus'*

In the year 1835-37, Purkinje and Mohi independently discovered that protoplasm is an important constituent of every cell and it plays an important role in every cell activity including division. Golgi (1838) discovered the golgi apparatus. Balbiani (1881) discovered chromosomes in the salivary glands of *Chironomus.* At about the same time, Flemming (1882) studied cell division in detail and gave the name *Mitosis.*

Endoplasmic reticulum was discovered by Porter in 1945, while Benda gave the name mitochondria to organelles originally discovered by Hemming. Lysosomes were discovered in 1955 by de Duve.

Shape of Cell

The shape of the cell may be variable as in the case of *Amoeba* or fixed. When the shape is fixed or permanent, all conceivable shapes are found-they may be spherical, rectangular, flattened, polygonal, oval, triangular, cone like, columnar etc. As in *Acetabularia,* the body of the plant made of a single cell is differentiated into a rhizoid like base, a long and slender stalk and an umbrella like cap. Certain cells, like the Ascospores of *Claviceps* are acicular (needle like).

Size

There is a great range of variation among cells in size also. Some of them are not visible to the naked eye, some are barely visible, while some are macroscopic This smallest cell size can be encountered in *Coccus* bacteria (0.2 to 0.5m), while the largest size of cell is seen in Ostrich egg (nearly 15 cms). The table given below summarises the variations in cell size.

Table 1 (Adopted from U. Sinha and S. Sinha, 1976)

Cell type	Size
Fibre cell of *Linum*	550mm (length)
Egg of ostrich	170x35mm
Egg of Hen	60x45mm
Human egg	0.1mm=100μm
Amoeba	100μ
Sea urchin egg	70μ
RBC	7μ
Typhoid bacillus	2.4 x 0.5μ

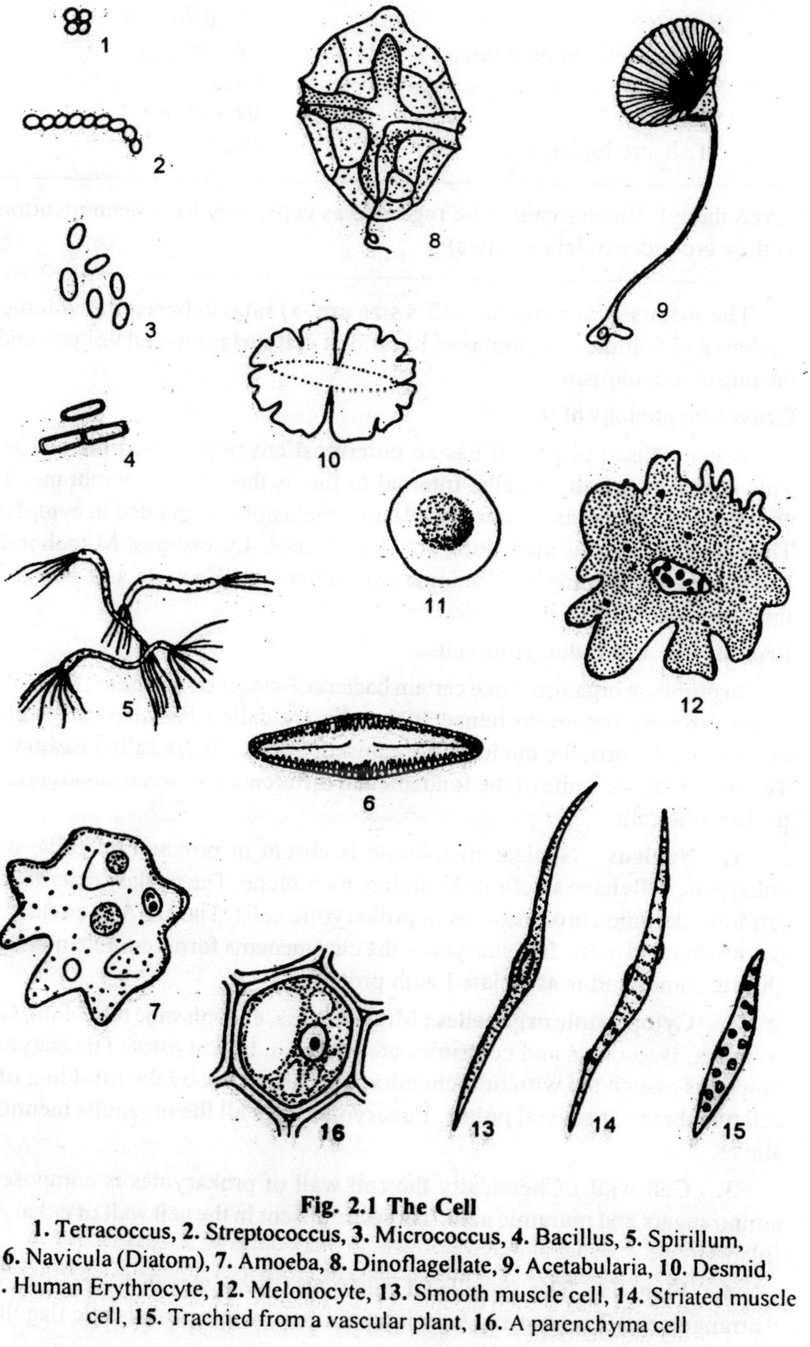

Fig. 2.1 The Cell

1. Tetracoccus, **2.** Streptococcus, **3.** Micrococcus, **4.** Bacillus, **5.** Spirillum, **6.** Navicula (Diatom), **7.** Amoeba, **8.** Dinoflagellate, **9.** Acetabularia, **10.** Desmid, **11.** Human Erythrocyte, **12.** Melonocyte, **13.** Smooth muscle cell, **14.** Striated muscle cell, **15.** Trachied from a vascular plant, **16.** A parenchyma cell

E. Coli	1.5 x 0.7μ
Diplococcus pneumonieae	200 x 100nm
*Influenze virus	100nm
*TMV	300 x 15nm
*T_3 bacteriophage	45nm,

(Even though Viruses cannot be regarded as cells, they have been mentioned as they are independent entities).

The factors governing the cell's size are -a) ratio between the volume of nucleus and volume of cytoplasm, b) ratio of cell surface to cell volume and c) the rate of metabolism.

Gross Morphology of the cell

A generalised plant cell has an outermost envelope called the cell wall. This is absent in animal cells. Internal to this is the plasma membrane. This encloses the nucleus and other cytoplasmic inclusions suspended in cytoplasm. The chief cytoplasmic inclusions are -Ribosomes, Lysosomes, Mitochondria, Plastids, Golgi complex, Endoplasmic reticulum, Vacuole and non living inclusions like crystals, raphides etc.

Prokaryotic and Eukaryotic cells

In primitive organisms like certain bacteria, blue green algae etc., the nucleus is not properly organised hence such cells are called *Prokaryotic,* while in evolved organisms, the nucleus is organised. Such cells are called *Eukaryotic.* The following are some of the fundamental differences between eukaryotic and prokaryotic cells.

1. Nucleus : Nuclear membrane is absent in prokaryotic cells, while eukaryotic cells have a definite bounding membrane. The nuclear material does not form definite chromosomes in prokaryotic cells. The DNA is circular and lies in a tangled mass. In Eukaryotes, the chromenema forms definite number of chromosomes and is associated with proteins.

2. Cytoplasmic organelles : Mitochondria, endoplasmic reticulum, Golgi complex, lysosomes and centrioles are absent in prokaryotes. The enzymatic functions associated with mitochondria are carried out by the infolding of the cell membrane at several points. Eukaryotes have all the organells mentioned above.

3. Cell wall : Chemically the cell wall of prokaryotes is composed of amino sugars and muramic acid. These are absent in the cell wall of eukaryotes (plants).

4. Flagella : The flagella of eukaryotes have eleven fibres with (2 + 9 arrangement) 2 central and 9 peripheral fibres, while prokaryotic flagella do not exhibit this arrangement.

5. Plastids : These are absent in prokaryotes. In photosynthetic prokaryotes the chlorphyll is associated with the lamellae, but the lamellae are not enclosed by any membrane. In contrast to this, in all eukaryotes, the chlorophyll pigment is found in definite organelles called chloroplasts.

In addition to the above differences, the streaming movement of cytoplasm may also be added, which is present in eukaryotes, but absent in prokaryotes.

Table 2. Comparison between Prokaryotic and Eukaryotic Cells
(Adopted from De Robertis Et al 1980)

	Character	Prokaryotic	Eukaryotic Cells
1.	Nuclear membrane	Absent	Present
2.	DNA	Naked and circular	Combined with proteins
3.	Chromosomes	Single	Multiple
4.	Nucleolous	Absent	Present
5.	Cell division	amitosis (fisson)	Mitosis or meiosis
6.	Ribosomes	70s (50s + 30s)	80s (60s + 40s)
7.	Endomembrane system	Absent	Present
8.	Mitochondria	Definite organelle absent enzymes of respiration and photosynthesis stored in the folds of plasma membrane	
9.	Choloroplast	Absent	Present
10.	Cell wall	Non cellulosic (present in plants only)	Cellulosic
11.	Exocytosis and endocytosis	Absent	Present
12.	Flagella	No definite arrangement of fibrills	9 + 2 fibrillar arrangement

PROKARYOTIC CELL

The prokaryotic cells are generally smaller and vary in size in different members. In mycoplasma it is about 0.12)μm while in *Oscillatoira,* a filamentous blue green the size is 40 x 5\μ. A great majority of them, however are about μ in size. Among the bacteria the smallest are to be found among cocci (0.1 μ) while the largest are the spirilla (60 x 6~m). The various components of the prokaryotic cell are described below.

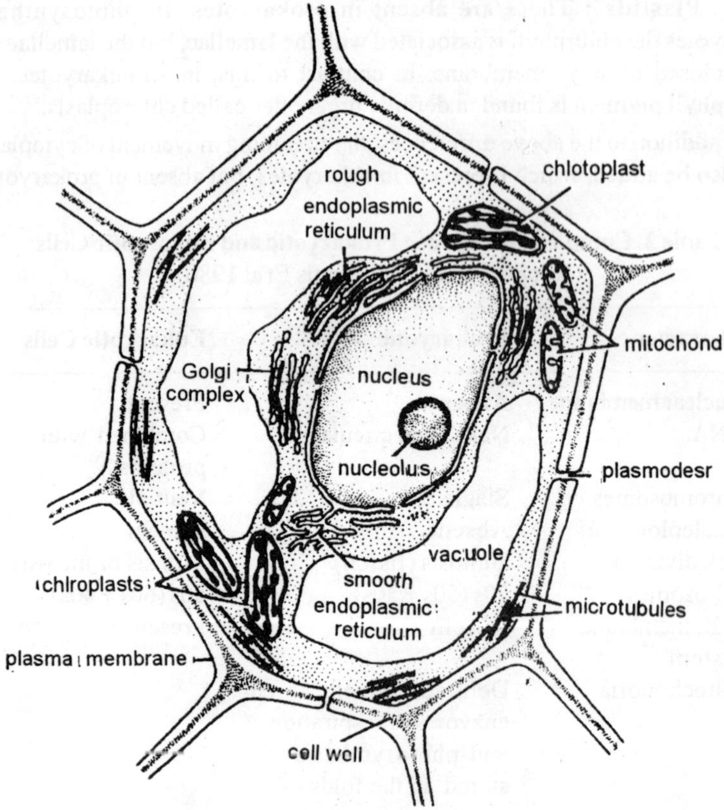

Fig. 2.2 The Cell
Structure of a typical Plant Cell (Electron Microscopic view)

Cell wall : Cell wall is absent in some prokaryotic (mycoplasma, L forms of bacteria) members while it forms the outermost boundary in a large majority of them. While the cell wall is non living and does not have the property of selective permeability it can act as a seive as it is porous.

In some prokaryotes the space between the cell membrane and cell wall is filled with some enzymes and metabolites consituting the periplasm.

Chemically the cell wall is made up of strong fibres of hetropolymers called *mucopeptides ofpeptidogycans*. They are also refered to as *glycopeptide, muropeptide, murein etc.*

The peptidoglycans consist of alternating units made up of N acetyl glucosamine and N acetyl muramic acid with 1-4 linkages. This type of a linkage is common to all Prokaryotes. Attached to the muramic acid is a short peptide chain consisting of alanine, glutamic acid and either lysine or diaminopimelic acid (DAP) and muramic acid. DAP and some of D-amino acids are found only in prokaryotic cell walls.

Cross linkage of these polymers assures structural rigidity with the type and extent of linkage being species specific. For example in some bacteria a pentaglycine links the tetrapeptide side chains extending from the muramic acid units while in others the terminal D- alanine of one tetrapeptide may be covalently linked to the DAP of an adjacent tetrapeptide. The peptidoglycans of gram negative bacteria have more cross linkage than those of gram positive bacteria.

In addition to peptiodglycans the wall has many other chemicals. The wall of gram negative bacteria shows as many as five layers under the electron microscope. These are-the periplasm space, Peptidoglycan layer, an intermediate zone, the outer membrane and lipopolysaccharides.

Cell membrane : This is the bounding layer of cytoplasm and is about 6-8nm thick. Unlike in the case of eukaryotic cells it carries out multiple functions. Some of these are -

(i) Presence of enzymes of biosynthetic pathways leading to the synthesis of wall components

(ii) Occurrence of *permease* enzymes responsible for the transport of materials across the membrane

(iii) Membranes of aerobic bacteria have the electron transport chain and oxidative phosohorylation

(iv) The purple bacteria have their photosynthetic apparatus in the membrane

(v) The membrane has attachment sites for chromosomal DNA (also plasmid) during DNA partitioning

(vi) Regulatory mechanism for chemotaxis is located in the membrane.

Structurally the membrane is triple layered consisting of two electron dense layers surrounding an electrontranslucent layer. In otherwords it is similar to a unit membrane. The architecture and chemical coposition is similar to the fluid mosaic model proposed by Singer and Nicolson (1972) (For details see later in the chapter).

Cytoplasm : Cytoplasm does not appear to have an organized structure when observed under electron microscope. It may be regarded as a concentrated solution containing a host of metabolites, enzymes, amino acids, inorganic ions, sugars, vitamins etc. Cytoplasm plays an important role in intermediary metabolism. There is no cytoskelcton.

Flagella (see bacteria)

Capsule (see bacteria)

Pili (see bacteria)

Surface layers : These are present in all gram negative and gram positive bacteria. Three main types of S layers have been reported in prokaryotic cells (bacteria) The S layers are composed of single homogenous peptides with occasional carbohydrates as minor components. These layers are believed to act as macromolecular sieves preventing the entry of toxic molecules and also to protect the peptidoglycans of the wall from the action of lytic enzymes.

Nucleoid : This represents the genetic material of prokaryotic cells. It is similar in all prokaryotes. (for details see - bacteria) The molecular structure of DNA is same as in eukaryotic cells.

Inclusion bodies : These are storage granules and other particles distributed in the cytoplasm. There are two kinds of inclusion bodies-membrane enclosed and membraneless (Shiviey 1974).

Membrane enclosed bodies : These are enclosed by a non unit type of membrane 2-4nm thick and made up of proteins. The inclusion bodies encountered in the prokaryotic cells are chlorosomes, carboxysomes, magnetosomes, gas vacuoles, polyhydroxybutyrate (PHB) granules, sulphur granules and glycogen granules.

PHB granules occur in cyanobacteria, sulphur granules are found in Thiorhodaccae and glycogen granules in *Clostridium* spp.

Chlorosomes have been demonstrated in photosynthetic bacteria belonging to chlorlobiaceae. (Cruden and Stanier, 1970). These are different from chromatophores in being non-membranous.

Carboxysomes are seen in cyanobacteria (Murphy et al 1974).

Magnetosomes are particles which help the bacterium in orienting itself in relation to the magnetic behaviour of the surroundings. Reported to occur in *Aquaspirillum magnetotacticum* (Balkwill et al 1980), the magnetosomes allow the bacteria to punise the most efficient aerotactic behaviour.

Non membranous Inclusion bodies : Phycobilisomes, cyanophycin granules (found in cyanobacteria), polyglucoside granules and several crystalline substances belong to this group.

In addition to the above there are R - bodies (retractile stuctures) demonstrated in bacteria by Lalucat and Mayer (1978). These are supposed to be similar to kappa particles of *Paramecium*.

Ribosomes (see bacteria)

Mesosomes (see bacteria)

Chromatophores : These are present in the photosynthetic bacteria belonging to the families Rhodospirillaceae, Chromatiaceae and Cyanophyceae. Chromatophores vary in form and have vesicles, tubes, bundles, stacks or thylakoids etc.

EUKARYOTIC CELL

Cell Wall

The cell wall is the characteristic feature of the plant cells. It is made up of non living matter and is believed to be secreted by the plasma membrane. The structure and composition of the cell wall not only varies in different plants, but in different types of cells of the same plant. Often, there might be variety in the cell wall composition at different stages of growth of the cell. The cell wall is very rigid and maintains the shape of the cell acting as a boundary. A unique feature of the cell wall is its permeability which is due to the presence of small pores.

Gross structure

Three layers are distinguishable in the cell wall. these are the *primary cell wall, middle lamella* and the *secondary cell wall.* The primary cell walls of adjacent cells are separated by middle lamella. Secondary wall is deposited after a certain period of development.

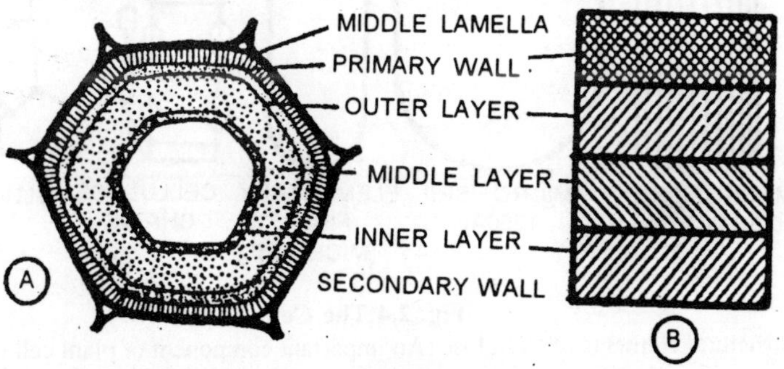

Fig. 2.3 The Cell
Arrangement of wall layers. **A.** Cross section, **B.** Arrangement of fibrils

The primary wall is secreted by the plasma membrane. In some instances like the leaves, fleshy stems, roots etc., the cells have only the primary wall and the middle lamella. The primary wall is formed during the early stages of growth and is about 1 to 3nm in thickness. Chemically the primary wall is composed of cellulose, hemicellulose and pectic compounds. The young primary wall is plastic and is prone to changes. Often it may be lignified.

The Middle lamella is the intercellular substance binding together the primary and the secondary walls. The middle lamella is formed during cell division and is amorphous and colloidal in nature. Chemically the middle lamella is composed of pectin, cellulose, calcium and magnesium pectate.

The secondary wall is deposited on the primary wall after cell maturation. It is very rigid and does not alter its shpae. The thickness varies from 5 - 10nm. The secondary wall usually has three layers viz., *outer layer, middle layer* and *inner layer.*

Chemically the Secondary wall is composed of cellulose, non cellulosic polysaccharides and hemicelluloses.

Ocassionally a *tertiary cell wall* may also be found as in the tracheids of some gymnosperms. This is a very thin layer and is composed of *xylan,* instead of cellulose. The secondary wall acquires different types of thickenings.

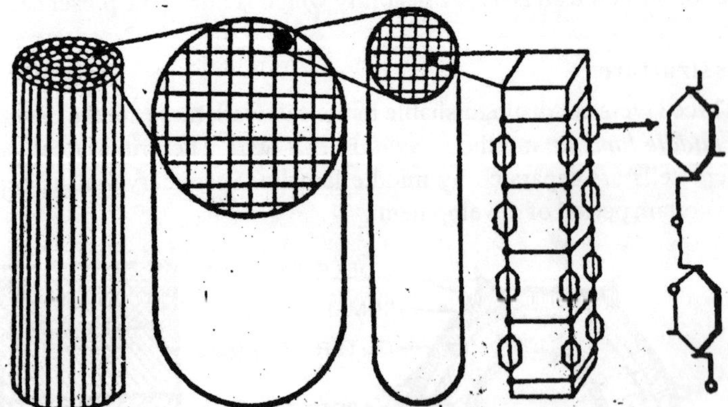

Fig. 2.4 The Cell

Structural elements of Cellulose (An important component of plant cell walls)

Ultra structure of the cell wall

The cell wall is mainly composed of cellulose. Cellulose is a polymer of glucose molecules. Approximately each molecule of cellulose consists of about 3000 glucose residues joined together by 1-4 oxygen bridges. These cellulose chains are oriented along the axis of the cell. About 100 of these cellulose chains bundle together to form *micellels* or *elementary fibrils.* Each micelle is about 100Å in diameter. Many hundreds of these micelles of these stack together to form *microfibrils.* Each microfibril is 200Å to 300Å in diameter. The micorfibril is visible only under eletron microscope. About 2000 of these microfibrils join together to constitute the *macrofibril* which is about 0.5μ in width. The macrofibrils are visible under the light microscope.

In secondary walls, the microfibrils are more orderly and compact than in the primary walls. The space between fibrils is filled by an interfibrillar substance usually the matrix. The matrix is filled by non cellulosic polysaccharides - such as pectin, hemicellulose, xylans etc.

The primary cell wall of dicotyledons contains many complex polysaccharides such as *xyloglucans, arabinogalactans* and *rhamnogalacturonans*. These are linked to the cellulose fibril and to one another by covalent linkages.

In certain walls, a glycoprotein called extensin is present. It is similar to collagen in being rich in 4 - hydroxyproline.

Extensin has a long protein chain, to the hydroxy groups of which are linked, residues of *arabinose* and *galactose.*

Plasmodesmata : Protoplasts of adjacent cells are connected to one another by means of thin cytoplasmic threads called plasmodesmata. These pass through the small pores (pit fields) present in the cell wall. Tubules which are in continuity in the endoplasmic reticulum are seen within plamodesmata thus maintaining an effective intercellular communication.

Origin of cell wall : At the completion of *Karyokinesis,* small vesicles of endoplasmic reliculum *(phragmoplast)* accumulate along the equatorial line to form the cell plate. The cell plate becomes thick by the addition of material from the golgi complex. Plasmodesmata are left open in the cell plate to maintain communication. This cell plate forms the lamella. Later, on either side of the middle lamella, cytoplasmic fibrils start accumulating to form the wall.

Plasma membrane : The cytoplasm in all the cells (both plant and animal) is surrounded by a living membrane called the *plasmalemma* or *plasma membrane.* This membrane is vital to the cell as it controls the exit and entry of all substances to and from the cell. The plasma membrane, as demonstrated by McAlear and Edwards (1958) in fungi, invaginates and maintains connection with the nucleus. It is also believed that many of the cell and their bounding membrane is same as the plasma membrane These organelles are nucleus, mitochondria, chloroplasts etc.

The presence of the cell membrane is proved by the following evidences.

1. On injury to a cell, when protoplasm flows out, it remains as a discreet unit.
2. Observations under electron microscope have revealed the presence of a boundary membrane over the surface of cytoplasm.

Size of the membrane : It is about 75Å thick and is composed of three concentric layers. The middle layer is about 35Å in thickness, while the outer layers on either side or 20Å thick.

Chemical composition : Membranes from different sources show variations in chemical composition. Essentially however all membranes are lipoproteinanceous. The components are held together in a thin sheet by non convalent bonds. The ratio of protein to lipids varies depending on the type (Golgi membrane, plasma membrane etc) of membrane, type of organism (eukaryote, prokaryote, etc.) and the type of cell (epidermis, phoelm etc.). In adition to these two, certain membranes also have carbohydrates.

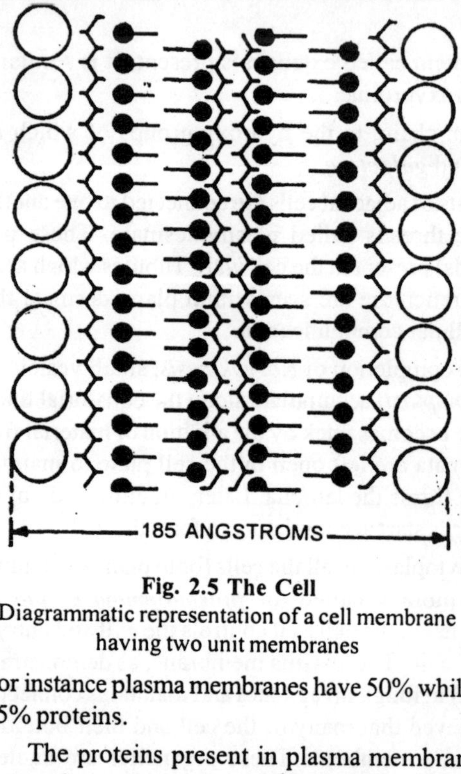

Fig. 2.5 The Cell
Diagrammatic representation of a cell membrane having two unit membranes

Membranes have three dfferent categories of proteins. These are - structural *proteins, functional proteins* (enzymes) and *carrier proteins.* Structural proteins maintain the structural frame work of the membranes. The average molecular weight of structural proteins found in membranes in $3x10^4$. Enzymes catalyzing various physiological activities are found in the membrane of mitochondria, endoplasmic reticulum etc. *Carrier proteins* are known to form a complex with the substances and help them to pass through the membrane.

The protein composition of different membranes vary (as has been pointed out already). For instance plasma membranes have 50% while mitochondrial membranes have 75% proteins.

The proteins present in plasma membrane are categorized into two viz. *integral proteins* and *peripheral proteins.* While integral proteins form the back bone of the membrane, peripheral proteins have a weak association maintained by weak electrostatic interaction. The integral proteins are hydrophobic in nature as they have to penetrate the hydrophobic lipids to form the basic structure of the membrane. These proteins do not easily disassociate from the membrane and it requires harsh chemical procedures to accomplish it.

The integral proteins have a hydrophilic portion also which protrudes out of lhe lipoid bilayer. Peripheral proteins are associated either with the hydrophilic head of lipids or hydrophilic part of the integral proteins.

The Lipids of membranes are of several types, but all of them share a common character in being *amphipatic* i.e. having both *hydrophilic* and *hydrophobic* portions within a molecule. There are three categories of membrane lipids-they are phosopholipids, glycolipids and sterols. While most membrane lipids are phospholipids, cholesterol does not have phosphates.

Phospholipids are of two categories viz. *phosphoglycerides* and *sphingophospholipids*. Phosplioglycerides have a glycerol back bone, while sphingophospholpids have the back bone derived from sphingosine an aminoalchobol having a long hydrocarbon chain.

Glycolipids contain one or many monosaccharide units attached to the lipids based on *sphingenine* or *ceramide*. There are three kinds of glycolipids viz., *Cerebrosides, sulphatides* and *gangliosides*.

Cerebrosides contain a single monosaccharide unit (galactose or glucose). Sulphatides are sulphuric acid esters of cerebrosides. Gangliosides are glycosphingolipids having *sialic acid*.

Sterols are steroid alcohols and are of three types - cholesterols (present in animal tissues), phytosterols (present in plants) and ergosterols (present in eukaryotic microorganisms).

Among the glycerophospholipids *lecithin* and *cephalin* are most abundant in higher plants and animals.

Membrane functions

The following are the major functions of membranes.

1. **Compartmentalization:** Membranes constitute the living boundary for cytoplasm and cell contents. As unbroken sheets they divide the living mater into self sustaining units to effectively coordinate and regulate the activities.

2. **Regulation of movement:** In a biologically active cell many susbstances enter and many others go out of the cell. However, if the entry and exit is unregulated, it is deleterious to the cell. Hence plasma membrane regulates and decides which substances are to be allowed and which should not. This unique property of the cell membrane called selective permeability or differential permeability is responsible for most of the physiological activities of the cell.

3. **Communication:** All the membranes have certain special substances (on their surface) called receptor molecules which mediate and control the message from one to another. The receptor molecules vary in different membranes. These molecules combine with substances (ligands) of different types and generate signals (in the form of cyclic AMP) directed to the interior of the cell. The best studied ligands are the hormones and neurotransmitters.

4. Interaction between cells: It is the responsibility of the plasma membrane as the living boundary between cells to effectively coordinate actions between cells since an organism consists of differnt types of cells and many a time the actions have to be coordinated to achieve harmony.

5. Energy transformations : Membranes are intimately associated with the inter conversion of energy. For instance, in green plants the light energy has to be converted into chemical energy. Similarly the chemical energy stored in carbohydrates has to be converted into energy currency i.e. ATP molecules. The machinery for energy capture and conversion is hidden in the membranes of mictochondria and chloroplasts.

6. Locus for biochemical functions : Membranes provide the structural framework for the location of various enzymes. The scaffolding of the membrane interconnecting the various organelles helps in the coordination of activities. Such an effective coordination of activities is best illustrated in the mitochondrial membrane where the various components of the electron transport chain work together as they are located in an orderly fashion.

7. Other activities : Membranes also function in various other capacities such as providing mechanical strength, involving in cell movement secretion etc., and also acting as electrical insulators.

Structure of the membrane : The molecular structure of the membrane has been interpreted in various ways by different cell biologists in order account for their known function. Various models have been proposed for the molecular structure of the membrane. These fall into two basic categories, viz., bilayer models and micellar or subunit models. The bilayer models are believed to have protein and lipid to be arranged in the form of layers, while the subunit models assume the presence of a number of independent subunits. We shall discuss a few of these models. The chart below summarises the list of various models.

MEMBRANE MODELS

- Bilayer models
  - Lipid membrane Model
  - Sandwich Models
    - (a) Danielli Davson model
    - (b) Unit membrane model
    - (c) Greater membrane odel
    - (d) Kavanau's model
- Micellar models
  - Models in which protein penetrate into the lipid layer
    - (a) Benson's model
    - (b) Lenard and Singer' model
    - (c) Mosaic membrane model
    - (d) The fluid mosaic model

BILAYER MODELS

Lipid membrane model : Overtoil (1895) discovered that fat soluble subsatances easily pass through the membrane indicating their affinity to the lipids. Hober (1910) and Fricke (1925) found out that the low electrical conductivity of the Iving cell is due to the presence of lipids. The experiments of Mudd and Mudd (1931) in placing an oil droplet which adheres to the cell surface provided the presence of lipids in the cell membrane. It was Gorter and Grendel (1925) who suggested a structural model based on lipids. They studied the erythrocyte (RBC) membrane and concluded that it is composed of two layers of liqid molecules whose polar groups are situated on the outside of the membrane.

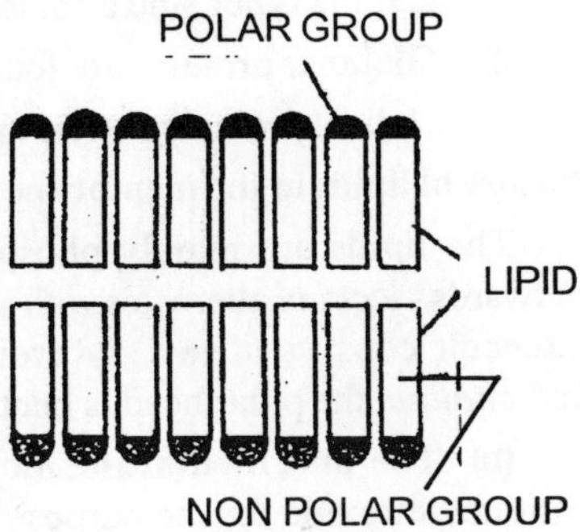

Fig. 2.6 The Cell
The lipid membrane model of Forter and Grendel

Sandwich models : In these models the lipid layers are supposed to be sandwiched between the two protein layers on either surface. The polar groups of the lipids adhere to the proteins on either side. The following are some of the models based on sandwich principle.

(a) Danielli Davson model : The study of surface tension of cells by Daniell and Harvey (1935) and Harvey and Cole (1931) indicated that there must be more to the structure of the membrane than merely being made up of lipids. In 1935, Hugh Davson and James Denielli proposed that the plasma membrane is a lipid layer sandwiched between layers of globular proteins. In a way the Danielli model was the first structural interpretation to account for the physiological properties of the membrane.

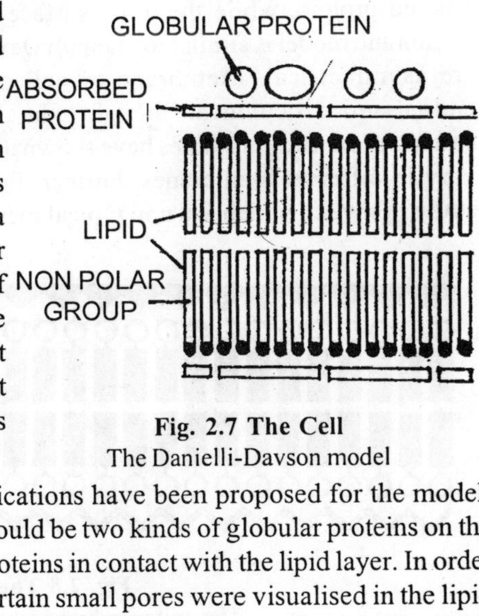

Fig. 2.7 The Cell
The Danielli-Davson model

Over the years various modifications have been proposed for the model. Danielli (1938) opined that there could be two kinds of globular proteins on the surface and diagonally arranged proteins in contact with the lipid layer. In order to account for the permeability, certain small pores were visualised in the lipid layers with an inner lining of protein molecules providing an inner hydrophilic surface.

The following are some other variations in the protein lipid arrangement.

1. Proteins are in the form of folded chain.
2. Proteins are in coiled from on both sides of the lipid layer.
3. Proteins are asymmetrically arranged i.e., one side has globular proteins and the other side a folded β chain.
4. Globular proteins are found on either side.
5. Helical proteins extend into the pores between the lipid molecules.

Nature of lipids in the membrane

The lipids are mainly phospholipids with their polar groups pointing outwards - *lecithin* alternates with a steroid molecule (*cholesterol*). Each lecithin molecule consists of two glycerol chains with the part containing *phosphate* and *choline,* the polar head is bent like a walking stick.

(b) The unit membrane model : The specimens of membrane fixed in osmium tetroxide for the purpose of electron microscopic observation reveal the occurrence of *two lines,* presumably indicating their double membrane nature. This observation originally made in 1950's by Humberto Fernandez Moran and David Robertson led to the unit membrane model. Robertson used a new fixative potassium permanganate. He discovered that each of the dense line possessed a sub structure having a trilaminar appearence. Each membrane is composed of two thin eletron dense lines of 20Å, with the intervening space of about 35Å. The inner and outer layers are non identical. The outer is a mucoid protein, while the inner surface is a non mucoid protein. The unit membrane model is similar to Danielli Davson model except that protein layers are asymmetrical. Membrane of cell organelles, plasma membrane and endoplasmic reticulum are thought to be similar in havingo the unit membrane structure. But current studies have shown that arrangement of lipids and proteins vary in different membranes. Further, the Robertson model is based on the study of myelin which is a non typical membrane.

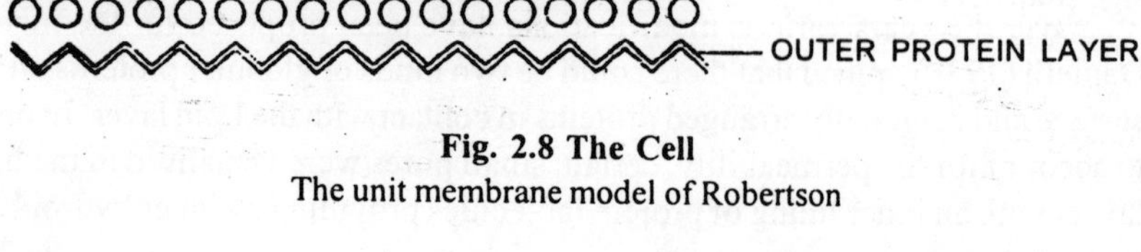

Fig. 2.8 The Cell

The unit membrane model of Robertson

(c) The Greater membrane model : This is similar to the trilaminar sandwich model except that the outer and inner surface proteins are different. The inner surface has an unconjugated structural protein adhering to polar heads of lipids. The outer surface also has structural proteins, on which is superimposed a *glycoprotein*. To this glycoprotein are attached side chains of oligosaccharides.

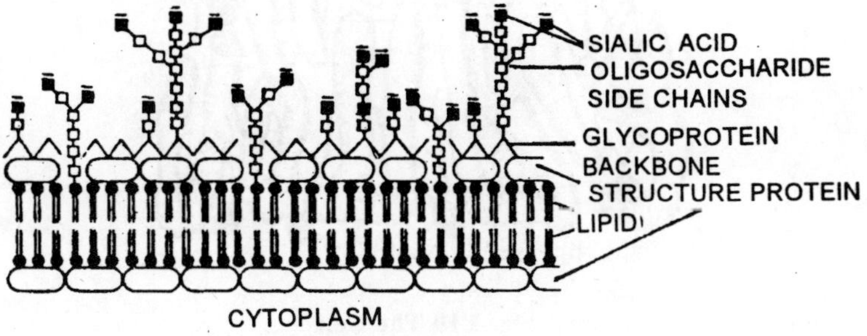

Fig. 2.9 The Cell
The greater membrane model

(d) Kavanau's lipid pillar model : According to Kavanau J.L. (1965), the lipid layer exists in two forms. Kavanau's model is essentially similar to Danielli Davson model. The two types of lipid structures are *pillars* and *flattened discs*. Pores are present between pillars to faciliate the passage of ions. Membranes with pillar lipids are thicker than those with discs. Internally the pillars have non polar tails, while the surface is made up of polar phospholipids. Proteins are present on either side of lipids.

Models based on protein penetration of lipids

(a) Lenard and Singer's model : This is based on the protein arrangement in the membrane. About 25 to 30% of proteins are helical while the rest are irregular coils.

(b) Benson's model : This model was proposed by Benson (1966) based on the study of membranes of chloroplasts. In this, it was postulated that lipid tails are bound to the corresponding hydrophobic points within the coils of proteins. The polar heads of the lipids lie on the surface.

(c) Mosaic membrane concept : According to this model proposed by Baum (1967), the membrane consists of units, each about 80Å in diameter. The wedge shaped protein is enclosed on either side and at edges by the lipids. The polar heads protrude out combining with the hydrophobic areas of proteins.

(d) The Fluid mosaic model : The molecular model of the plasma membrane proposed by Singer and Nicolson (1972) emphasises on the dynamism of membrane organisation and visualises it to be a semifluid like structure with a mosaic arrangement of lipids and proteins. The fluid mosaic model, though essentially based on bilayer models, rejects the idea of continuous bilayer.

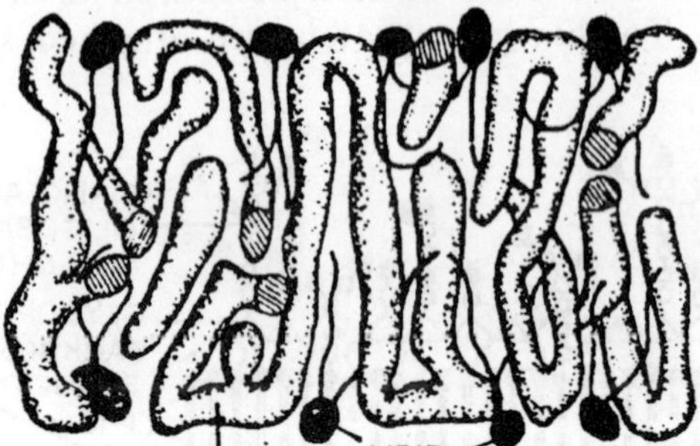

Fig. 2.10 The Cell
Membrane model of Benson

Instead of forming continuous external layer adhering to the hydrophilic heads of the phospholipids, as in the Danielli model; the proteins constitute discontinuous particles often penetrating deeply into the lipid layer. In a way the protein units could be compared to floating incebergs in a sea of phospholipids. The countinuity of the bilayer is not maintained because of complete or incomplete interruption by proteins. While the Danielli model assumes a hydrophilic bonding between lipids and proteins, the Singer Nicolson model regards it to be hydrophobic.

The membrane proteins in the fluid mosaic model are of two types distinguishable by their intensity of lipid association. These are the *integral* or *intrinsic proteins* and the *extrinsic* or *pheripheral proteins.* ITie integral proteins penetrate deep into the lipid layer and are the hydrophobic as they have to coexist in the hydrophobic environment of fatty acids. The integral proteins are *amphipatic.* While their tips are hydrophilic; the central portion is hydrophobic. These proteins are difficult to extract from the membranes.

The peripheral proteins which are weakly associated whith the membrane can be easily extracted. They are hydrophilic and never penetrate the interiors of the lipid bilayer. They form weak bonds either with the polar heads of lipids or with the hydrophilic surfaces of integral proteins. One of the best evidences for a broken surface layer of proteins comes with the using of phospholipases. These phosphate degarding enzymes very easily percolate into the membrane, which would have been impossible, if there were to be an unbroken surface envelope of proteins.

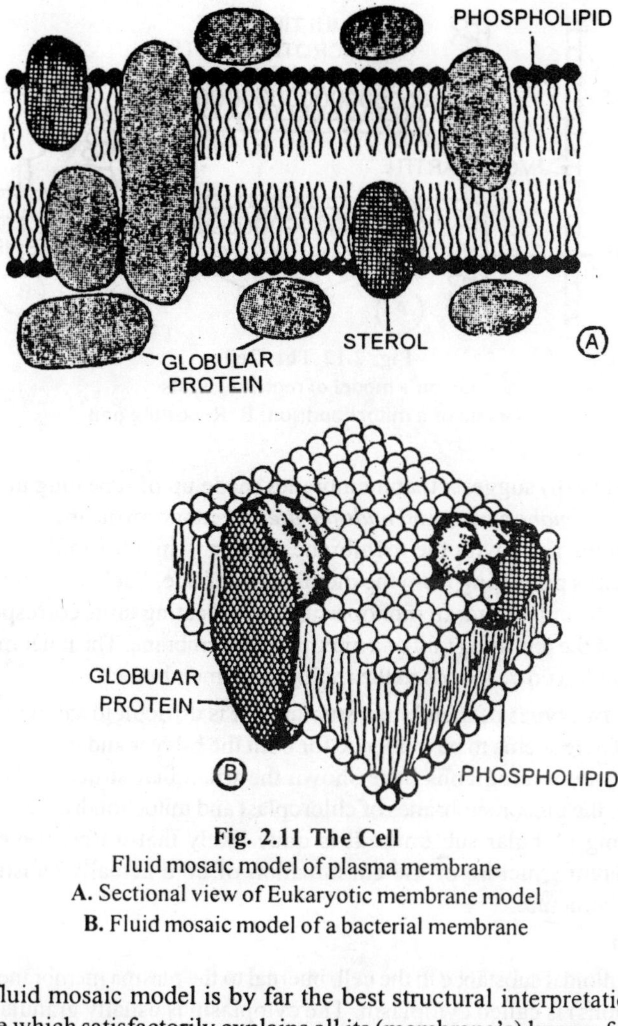

Fig. 2.11 The Cell
Fluid mosaic model of plasma membrane
A. Sectional view of Eukaryotic membrane model
B. Fluid mosaic model of a bacterial membrane

The fluid mosaic model is by far the best structural interpretation of the membrane which satisfactorily explains all its (membrane's) known functions.

MICELLARMODELS

According to Fernandez Moran (1963) the membrane is corpuscular in nature consisting many repeating units. These units called *globular units* or elementary particles range in diamiter form 40 - 70Å and form closely packed repeating units. The membrane structure may be globular or unit membrane in different organelles or under different metabolic conditions. According to some cell biologists, the corpusular nature is an artifact.

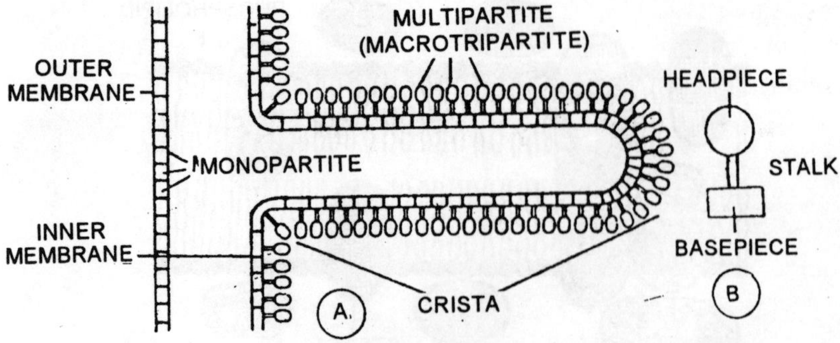

Fig. 2.12 The Cell
Green's model of repeating units
A. Crysta of a mitochondrion, **B.** Repeating unit

Green (1970) suggests that membranes made up of repeating units are of two types viz., *monopartite* and *multipartite* The monopartite membranes have repeating units flush with the membrane surface, while the multipartite have repeating units projecting from the membrane sufrace. Each repeating unit has a base, a stalk and a head. In monopartite, the repeating units correspond only to the base of the repeating unit of a multipartite membrane. The inner membrane of mitochondria corresponds to the multipartite membrane.

Of the two types of membrane stuctures, it is difficult to say which model is correct There seems to be evidence for both the bilayer and micellar models. While electron micrographs have shown the trilaminar structure for the unit membrane, the inner membranes of chloroplast and mitochondria have revealed the repeating globular sub units. It is quite likely that different membranes show different structure or the combination of both actually constitutes the membrane structure.

Cytoplasm

The colloidal substance in the cell, internal to the plasma membrane (without the inclusions) is called cytoplasm. The cytoplasm is usually granular and has many substances either dissolved or suspended in it The suspended particles may be in the solid or semi solid condition. There are several theories regarding the structure of the cytoplasm. Some of these are-

Reticular theory : Cytoplasm is made up of network of fibrils or particles.

Alveolar theory : According to Butschill (1892), cytoplasm consists of a number of droplets or alveoli. It gives an appearence of oil droplets floating on water.

Granular theory : According to Altmann (1893), Cytoplasm is composed of a number of minute granules suspended in a matrix.

Fibrillar theory : This theory was proposed by Fisher (1894) and Flemming (1897). Electron microscopic studies have proved the hypothesis of Fisher. According to this view, cytoplasm is made up of a number of fibrils.

Properties of cytoplasm : Cytoplasm has a number physical and biological proerties.

Physical properties

The colloidal nature of cytoplasm is responsible for the various physical properties A colloidal system may be defined as a system of liqid medium containing suspends particles varying in diameter from 1/1,000,000 to 1/10,000 nm. Some of the physical properties are follows.

Phase reversal : Cytoplasm may be differentiated into a *sol state* and *gel state.* Sol state is the liquid state and the gel state is the semi solid state. These two phases are interchangeable. This capacity is known as phase reversal.

Elasticity : Depending on the circumstances, the cytoplasm can extend or contract subject to a certain limit. This is known as elasticity.

Cohesiveness : The suspended particles in cytoplasm have mutual attraction, thus exhibiting cohesiveness.

Contractility : It is the capacity of peripheral cytoplasm to absorb or remove water from cell to the exterior. This is manifested very well by guard cells.

Viscosity : The suspended particles of cytoplasm are responsible for its viscous nature.

Brownian movement : The suspended particles of cytoplasm are in a state of to and fro movement called Brownian movement

In addition to this, cytoplasm also exhibits movements (seen in the plasmodium of slime molds) and 'cyclosis or streaming movements (seen in the leaf cells of *Elodea).*

Biological properties

Growth, respiration, excretion, nutrition, reproduction, metabolism, irritability etc., are some of the important biological activities of cytoplasm.

Chemical composition of cytoplasm : 90% of cytoplasm is water. The remaining 10% is constituted by organic and inorganic compounds.

About 40 or more inorganic constituents are found in cytoplasm. These are found in the form of ions (either cations or anions). Some of the common inorganic constituents are - Oxygen (62%), Hydrogen (10%), Carbon (18%) and Nitrogen (3%). The less common inorganic contituents are Calcium (0.25%), Phosphorus (1.2%), Sodium (0.1%), Magnesium (0.7%), Iodine (015%) and Chlorine (0.15%).

Among the various organic substances proteins, carbohydrates, lipids, hormones, vitamins and nucleic acids are most important.

Cytoplasmic Inclusions : Cytoplasm has a number of living and non living inclusions. The living inclusions are - Nucleus, mitochondria,. plastids, endoplasmic reticulum, golgi complex, lysosome, centrioles, microbodies, etc.

Nucleus

This is the most important organelle of the cell which regulates all its activities. It was discovered by Robert Brown in 1831.

Position : Usually it is present at the centre of the cell, but in some instances as in mature parenchyma cells, the nucleus is peripheral. In *Acetabularia,* the nucleus occupies various positions, though mainly present at the basal region.

Shape, size and number : The shape is variable depending on the cell type. Though the nucleus is generally spherical, ellipsoidal or flattened ones also may be seen in certain cells. The surface is generally smooth but as in leucocytes it may have small infoldings.

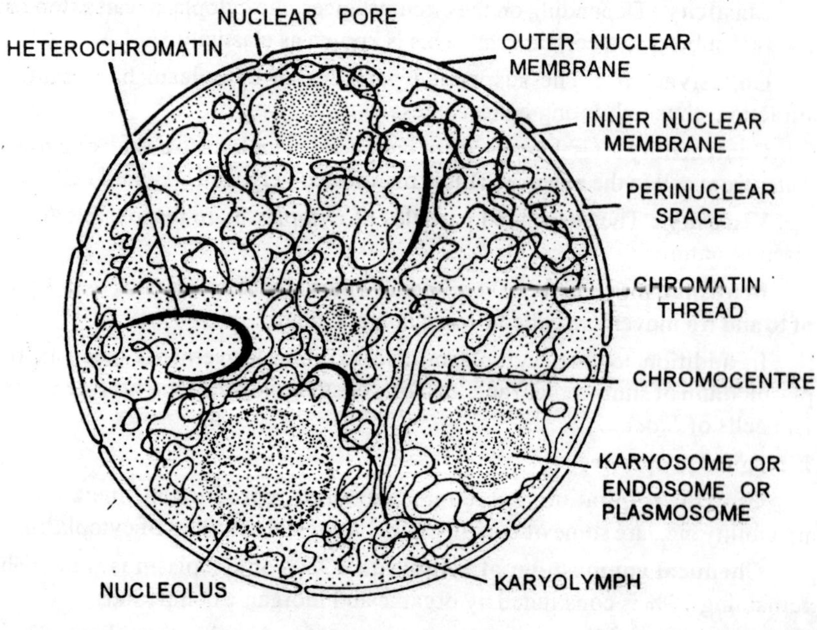

Fig. 2.13 The Cell
Structure of the nucleus

The size of the nucleus is also variable depending on the cell; the size of the nucleus being directly proprolional to that of cytoplasm. The following formula or Heywig may be used to determine the size of the nucleus in a particular cell.

$NP = Vn/Vc - Vn$

When NP = nucleoprotein index, Vn is the volume of the nucleus and Vc is the volume of the cell.

Most of the cells are uninucleate, while some *(Paramecium)* may be binucleate or multinucleate *(Cladophora)*. In certain fungi like *Mucor,* where there are no septa dividing the plant body into cells, numerous nuclei occur indicating a coenocytic condition.

Ultrastructure of the nucleus : Ultrastructurally the nucleus has the following parts - 1. Nuclear membrane, 2. Karyolymph or Nuclear sap, 3. Chromonemata, 4. Nucleolus and 5. Endosomes.

Nuclear membrane : In all eukaryotic organisms, a limiting membrane envelops the nucleus. This membrane helps in effective communication between nucleus and cytoplasm. The elements of endoplasmic reticulum contribute to the nuclear envelop during cell division.

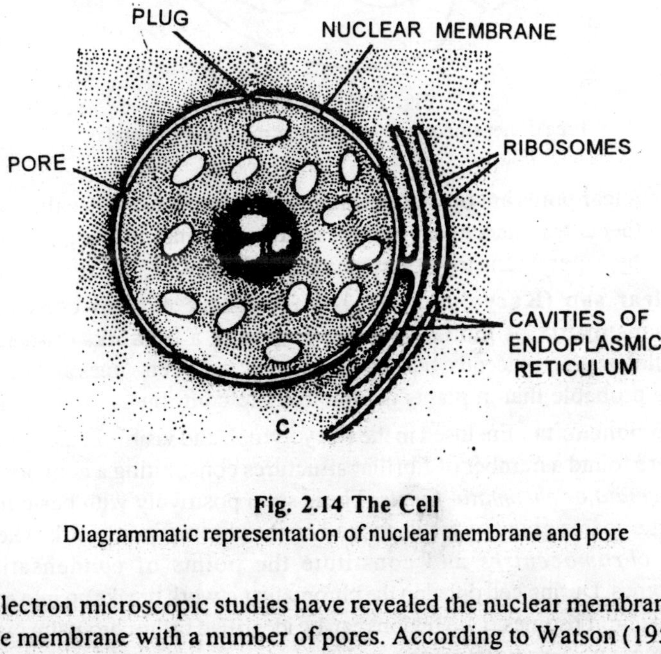

Fig. 2.14 The Cell
Diagrammatic representation of nuclear membrane and pore

Electron microscopic studies have revealed the nuclear membrane to be a double membrane with a number of pores. According to Watson (1959) these pores called *nuclear pores* are octogonal in shape. The outer nuclear membrane is called endokaryotheca. Each membrane is about 75 - 90Å thick and they enclose a space between them about 100 - 150Å (De Robertis et al 1970). This space is called the *perinuclear space* or *cisterna.* The outer membrane is rough due to the presence of RNA particles. Internally the nuclear membrane is followed by a supporting membrane called *fibrous lamina* (300Å thick).

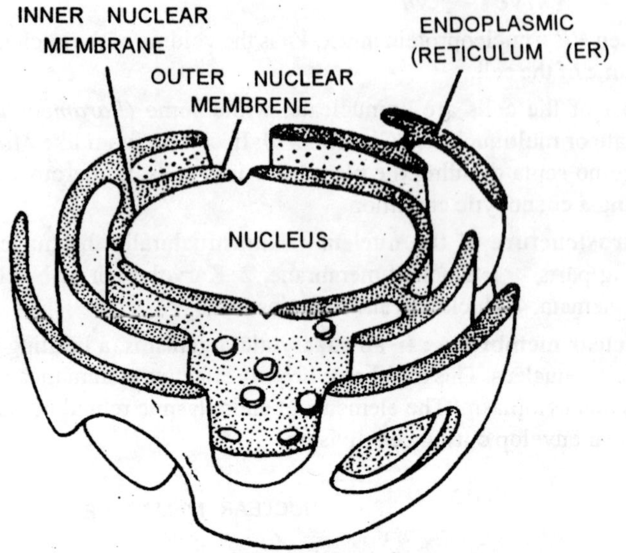

Fig. 2.15 The Cell
Three dimensional view of nuclear membrane complex

The nuclear pores are 600Å wide and are enclosed by rings called *annuli.* These together constitute the pore complex In some instances, the pore complex is covered by a membrane to regulate the premeability.

Nuclear sap (Karyolymph) : The interphase nucleus consists of a homogenous matrix of fluid substances enclosing a network. Primarily the Karyolymnh is proteinaceous, but also has nucleic acids, enzymes and minerals. It is quite probable that in plants the nuclear sap contributes to the spindles.

Chromonemata : Enclosed in the Karyolymph and visible in the interphase nucleus are found a number of fibrillar structures constituting a network called *chromonemata* or *chromatin fibrils.* These stain positively with basic fuchsin stain. Some coarse granules are deposited on the chromatin network. These are a called *chromocentres* and constitute the points of condensation of chromosomes. During cell division the chromatin network breaks up into specific number of chromosomes. Two regions can be identified in the chromatin material. These are *heterochromatic* region and *euchromatic* region.

The heterochromatic region stains darkly and shows numerous bead like structures called *chromomeres.* The heterochromatic region has less DNA. This region is believed to be genetically and metabolically inert. The light staining region of the chromatin is called the euchromatic region. This region contains more of DNA and is supposed lo be genetically active.

Nucleolus : This was first discovered by Fontana (1874). A spheroidal body, situated either in the central or pheripheral position, the nucleolus is supposed to regulate the synthetic activity of the nucleus. Usually two or more chromosomes are associated with the nucleolus (this can be seen during late prophase) and these are called *nucleolar organisers* as they play a role in the reappearence of the nucleolus after cell division. The number of nucleoli per nucleus varies form one to two or three.

Chemically the nucleolus is rich in RNA. According to Palade (1963), the nucleous of liver cells contain 3 - 5% RNA, while pea embroys have 10 - 20% of nuclear RNA. Orthophosphates helping in RNA synthesis have been reported in nucleolus. Besides this many enzymes (acid phosphatase, nucleoside phosphorylase, RNA methylase etc) have been reported in the nucleolus.

Types of nucleoli : According to Wilson, there are two kinds of nucleous viz., Plasmosomes and Karyosomes (Ocata). The plasmosomes are positive to acidic stains and have a transparent outer region with a dense interior. The karyosomes stain positively with basic dyes. Three kinds of karyosomes have been identified-*Netknot* type (like nodules attached to threads), *Chromosome nucleoli* (occuring only in gametocytes) and *Karyospheres* (spheriodal bodies with basic chromatin).

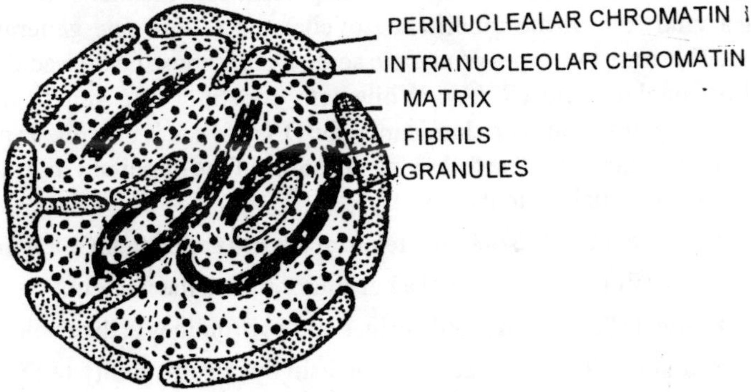

Fig. 2.16 The Cell
Schematic representation of Ultra structure of nucleus

In some instances, plasmosomes and karyosomes combine to form amphinucleoli; these are common in molluscan eggs.

Fine structure of the nucleolus : Detailed analysis of the nucleolus has revealed the presence of three regions. These are granular region, fibrillar region and amorphous region. The granular region occuring at the peripheral region of the nucleolus is composed of granules of 150-250Å thick. This region has ribonucleic acid and proteins. The fibrillar region also known as nucleolonema is composed of fibrils 50-80Å in length. The amorphous region has low electron density and is the one to disappear first at the time of cell division.

There is no limiting membrane to the nucleolus and the organisation is maintained by the calcium ions.

Functions of nucleolus

1. It is the active site of RNA synthesis
2. It is the source of ribosoinal RNA
3. It produces precursors of ribosomes
4. According to Maggio (1990), protein synthesis takes place in the nucleolus. But this may not be totally true.

Endosomes : These are granular structures present in the karyolymph and are smaller in size than the nucleolus.

Chemical composition of Nucleus : The main constituents of nucleus are nucleoprolcins, besides enzymes and inorganic salts. The following gives the total distribution of nucleoproteins in the mammalian liver cells-DNA 9%, RNA 1%, Histones 11%, Residual proteins 14%, and other acidic proteins 65%.

Nucleic Acids : These are the most important constituents of the nucleus playing a vilal role in the inheritance of characters from one generation to-the other. There are two types of nucleic acids-Deoxyribose nucleic acid (DNA) and Ribose nucleic acid (RNA). While both constitute the genetic material in different organisms, it is predominantly the DNA which is the genetic material in most of the organisms. The following are some of the important milestones in the discovery of Nucleic acids.

1. Meischer (1869) isolated nucleic acids and called them nuclein.
2. Hertwig (1884) suggested a heriditary role to the nuclein.
3. Kossel (1884) recognised purines and pyrimidines in the nucleic acids.
4. Feulgen (1912) devised basic fuchsin dye positive only to DNA.
5. Astbury and Bell (1938) showed that the bases in DNA molecule are arranged one above the other.
6. Avery, Macleod and Macarty (1944) presented evidence for DNA to be the genetic material.
7. Chargaff (1950) showed that DNA contains equal amount of puriness and pyrimidines.
8. Wilkins et al (1950) identified the molecular distances between the bases.
9. Furberg (1952) opined that DNA is a coil of single chain.
10. Pauling and Corey (1952) suggested that DNA is a helix of three chains.

11. Waston and Crick (1953) demonstrated the double helix model for DNA molecule.
12. Meselson and Stahl (1958) demonstrated the semi conservative replication of DNA molecule.
13. Jacob and Monad (1961) postulated the presence of RNA in protein

Deoxy Ribose nucleic Acid (DNA) DNA is the main genetic material in almost all organisms except for certain Viruses.

In prokaryotes like *Escherichia* coil the genetic material is a single giant molecule of DNA. While DNA is mainly concentrated in the nucleus, it has also been reported in cytoplasmic organelles like chloroplast, mitochondria etc.

Chemical composition of DNA : DNA extract from many living organisms has revealed it to be a complex macromolecule. It is a long chain polymer composed of monomers (units) called *nucleotides.* Each nucleotide in turn is composed of three sub units. These are a) nitrogen base, b) sugar and c) phosphate.

Nitrogen bases : The nitrogen bases are of two kids viz. *pyrimidines* and *purines.* Both pyrimidines and purines are heterocyclic compounds, with a basic six membered ring in which one or the other carbon position is replaced by nitrogen.

The pyrimidines have a basic six membered ring of benzene in which, one or the other carbon position is replaced by nitrogen. There are two kinds of pyrimidines-*cytosine* and *thymine.* Structurally cytosine has a single six membered ring with a hydroxyl group at the 2' position and an amino group at the 6' position (*2 hydroxy, 6 amino pyrimidine*). Thymine also has a single six membered ring with hydroxyl groups at the 2' and 6' position .and a methyl group attached to the 5' position (*2.6' hydroxy, 5 methyl pyrimidine*).

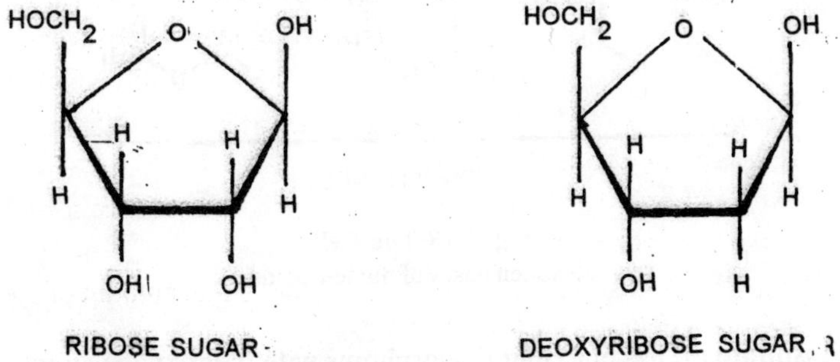

Fig. 2.17 The Cell

Pentose sugars of nucleic acids (left, robose; right deoxyribose)

The purines have a double ringed structure. Attached to the six membered benzene ring is a 5 membered imidazole ring. The imidazole ring is joined to the hexagonal ring at 4' and 5' positions. The benzene ring has nitrogen in the 1' and 3' position (replacing C).

There are two types of purines viz., *Adenine* and *Guanine*. Adenine has an amino group at the 6' position (*6 aminopurine*), while guanine has an aminogroup at the 2' position and a hydroxyl group at the 6' position (*2 amino, 6 hydroxy purine*).

Sugar : The sugar molecule in DNA is basically a ribose sugar. The ribose sugar is a pentose sugar. It is a 5 membered ring. The ring consists of four carbon atoms while thc 5th C atom is outside the ring. Position 1 '.3' and 5', in the ring have OH groups while the 2nd carbon position has only H. There is no O; hence the sugar is said to be Deoxyribose (no oxygen in the second carbon position).

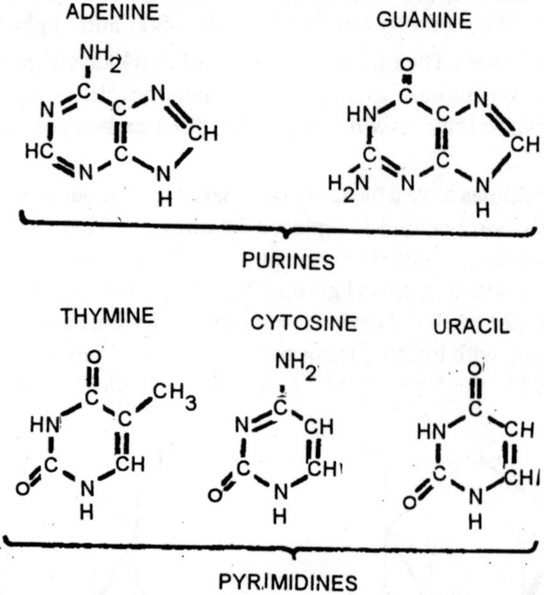

Fig. 2.18 The Cell
Nirtogen bases of nucleic acids

Phosphate : It has one central phosphorus with four oxygen atoms.

Nucleosides : A nitrogen base coupled with sugars forms a nucleoside. DNA has four nucleosides : a) Deoxycytidine, b) Deoxythymidine, c) Deoxyadenosine and d) deoxyguanosine. The sugar molecule is attached to the nitrogen base to the 3' position of the imidazole ring in the case of purines.

Nucleotides : A nucleotide is derived from the nucleoside by the addition of phosphoric acid molecule to the latter. The phosphate group is usually linked to the two sugar molecules above and below. The linkage is with the 3rd C atom of one deoxyribose sugar and with the 5th C atom of another. DNA has four nucleotides- Deoxycytidilic acid, Deoxy thymidilic acid, Deoxy adeniylic acid and Deaxyguanylic acid.

Polynucleotide : DNA is said to be a polynucleotide chain with a number of nucleotides joining together with the help of DNA. polymerase.

Molecular model of DNA structure : A molecular model of DNA structure first proposed by J.D. Waston and F.H.C. Crick (1953) has been universally accepted. They were awarded Nobel prize for their discovery. The work of Crick and Watson has been supported by X-ray diffraction picture obtained by Wilkins and his associates.

Fig. 2.19 The Cell
F.H.C. Crick (Left) and J.D. Watson (Right)

According to this model, a DNA molecule is composed of two chains that are spirally coiled and form a double helix, much like a spiral stairway. The distance between the two chains is uniform and it is maintained by base pairing. The back bone of the chain is made up of alternating sugar and phosphate molecules. At specific , intervals, the chains are joined by 'steps.' Each step is made up of two nitrogen bases. Each nitrogen base is attached to the sugar

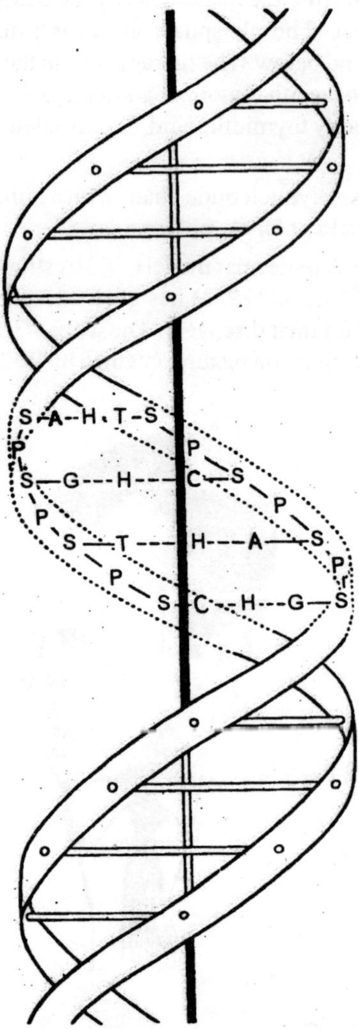

Fig. 2.20 The Cell
The double helix model of DNA

molecule of the opposite chain. Between themselves, the nitrogen bases of a 'step' are joined by hydrogen bonding. The pairing of the nitrogen bases is always specific. It is made up of a purine and a pyrimidine i.e., a double ring and a single ring. Two purines would occupy too much space and can not be accommodated between two chains: similarly two pyrimidines would form a short step and leave a gap between two chains. Hence the pairing is always between a purine and pyrimidine. Because of this pairing, in a DNA molecule total purine content is always equivalent to total pyrimidine content

The purine pyrimidine base pairing is further specific. It is always Adenine with Thymine and Guanine with Cytosine. Adenine and Thymine are joined by two hydrogen bonds through atoms attached to positions 6' and 1' Cytosine and Guanine are joined by three hydrogen bonds at position 6', I' and 2'. The bonding of hydrogen is weak; but this facilities easy separation of two strands during replication. According to Crick and Waston, the helix has a diameter of 20Å and there is one complete turn of the helix at every 34Å and has a stack of 10 nucleotides at every turn.

One unique character of DNA is its unity and yet infinitesimal variety. The DNA structure mentioned above is uniform in all organisms, whether it is man or virus. At the same time it is also true that DNA is the genetic material that decides the characters of individual organisms. Then how does DNA decide the individuality. The answer lies in the base pair sequence. The combination of A/T, G/C is uniform, but their sequence is not; for instance in one individual it may be AT, GC, AT, AT, AT, in another-GC, GC, AT, GC, AT etc. There is no limit to the variety in which base pair sequences are arranged. This is the secret of individual differences. In no two individuals, the base pair sequence is same.

Unusual bases In DNA : The normal bases in DNA are A, T, G, and C. In some viruses like PBSIand PBS2 uracil replaces thymine. In some bacteriophages C is replaced by 5 hydroxy methyl cytosine.

Single strand DNA : Although DNA in most organisms is double stranded in some phage viruses (f x 174), single stranded DNA is found. The following table summarises the differences between double stranded DNA and single stranded DNA.

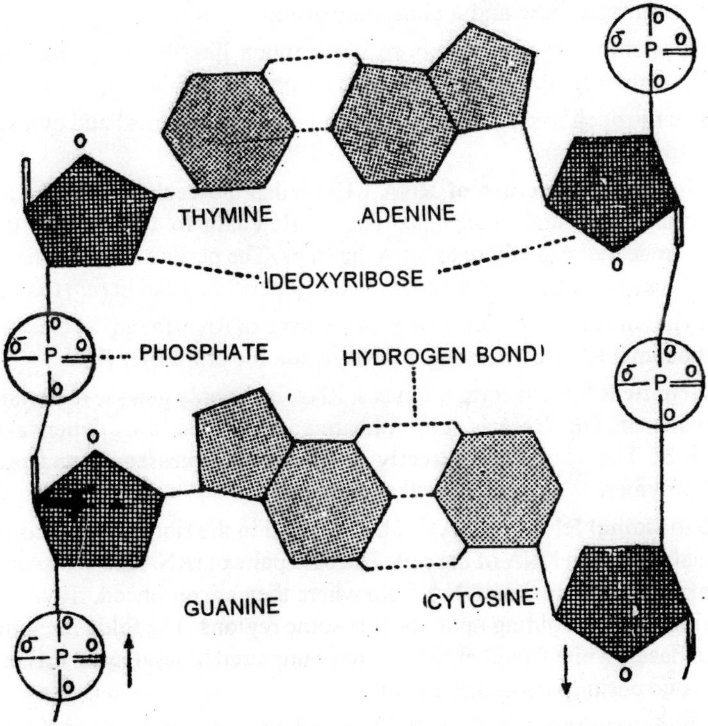

Fig. 2.21 The Cell
Arrangement of nucleotides in a segment of the DNA model

Table 3

	Character	Double Strand DNA	Single Strand DNA
1.	U.V. absorption	constant from 10-80°C and then rises rapidly	steady increase from 2° to 90°C.
2.	Action of formaldehyde	Resistant	not resistant
3.	Ratio of nucleotides	Equal, A = T and G = UC	A: T: G: C is 1.33:0:98:0.75
4.	Shape	Linear	Circular
5.	Replication	two strands separate and replication begins	becomes double stranded (replicate from) temporarily)

Ribonucleic acid (RNA) : The RNA is also a polymer of nucleotides. These are called ribonucleotides. Each ribonucleotide has three parts viz., a pentose sugar, a nitrogen base and a phosphate group.

The pentose sugar is a ribose sugar much like that of DNA, but with oxygen (present) attached to the 2' carbon position of the ring.

The nitrogen bases are - Adenine and guanine (purines) and cytosine and uracil, (pyrimidines)

Molecular structure of RNA : Here also the back bone is made up of alternating sugar and phosphate. The single chain folds to enable the base pairs to meet here and there to form the steps. The pairing is between A-U and C-G. Occasionally the RNA occurs as the genetic material in *reoviruses*.

Types of RNA : The following are the types of RNA found - a) Genetic RNA b) Ribosomal RNA c) messenger RNA d) soluble or transfer RNA.

Genetic RNA : In certain viruses, RNA is the only genetic material, DNA being absent. This RNA is self replicating; it is called RNA dependant RNA synthesis. The viral RNA directly functions as a messenger and produces RNA polymerase enzyme as well as proteins.

Ribosomal RNA (rRNA) : This is found in the ribosomes. It constitutes about 80% of total RNA of the cell. The base pairs of rRNA are complementary to that part of the nuclear DNA from where they are produced. rRNA consists of a single strand folding upon itself at some regions. The folded regions have paired bases, while the other regions have unpaired bases. Due to this in rRNA there is no purine pyrimidine equality.

There are three types of rRNA based on sedimentation and molecular weight. These are a) rRNA with molecular weight over a million, eg. 21s - 29s rRNA, b) rRNA with molecular weight less than a million, eg. 12s- 18s RNA, c) rRNA with very low molecular weight (40,000) eg 5s rRNA.

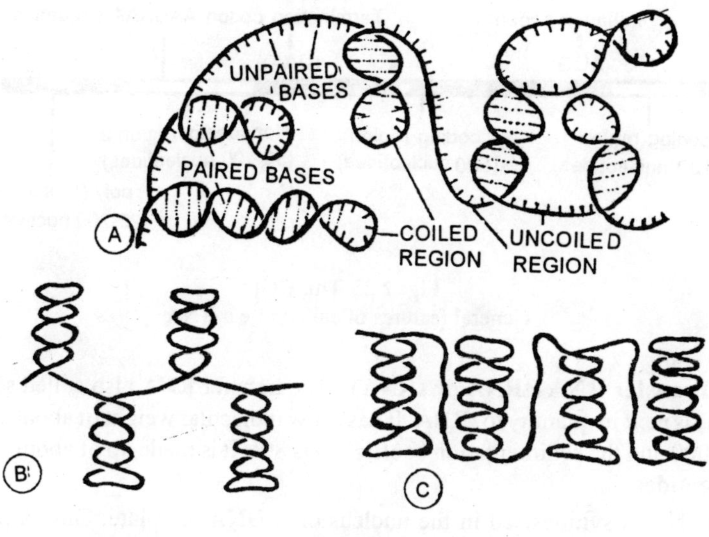

Fig. 2.22 The Cell
Secondary structure of ribosomal RNA (Diagrammatic)
A. Extended strand, **B.** Negative interaction showing repulsion,
C. Positive interaction showing association

Messenger RNA (mRNA) mRNA carries the message from DNA to ribosomes during protein syntheses. It is about 3 - 5% of total RNA content of the cell.

mRNA has a molecular weight of about 5,00,000 with a sedimentation coefficient of 8s. This however is highly variable. nRNA is always single stranded and there is no base pairing, even though random coiling may be observed. The mRNA molecule is produced from the coding strand and its base sequence is complementary to the coding strand. The following are some of the general features of mRNA.

1. **Cap :** This is a blocked methylated structure found at the 5' end of RNA.
2. **Non coding region :** This follows the cap and consists of 10 - 100 nucleotides. There are no codes for proteins in this region.
3. **Initiating codon :** This region begins the coding.
4. **Coding region :** Codes for proteins and has about 1500 nucleotides.
5. **Terminating codon :** The coding region is followed by a terminating codon, which stops translation.

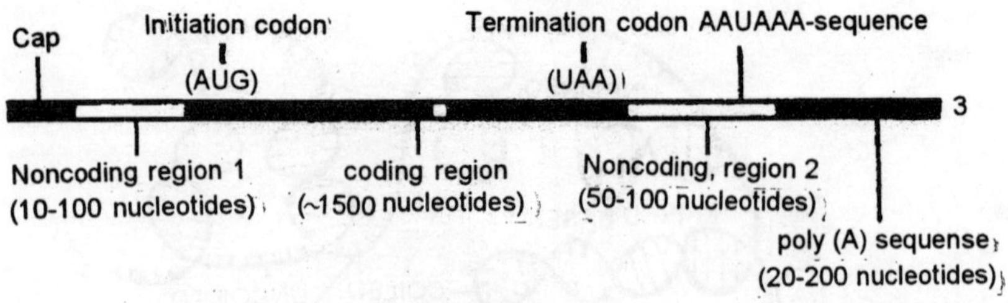

Fig. 2.23 The Cell
General features of eukaryotic mRNA

Transfer RNA (tRNA or sRNA) : The transfer RNA also called soluble RNA, is next in quantity to rRNA. It has a low molecular weight of about 25,000 - 30,000 and the sedimentation coefficient is 80s. It is made up of about 70 - 90 nucleotides.

tRNA is synthesised in the nucleus on a DNA template. Only a minute fraction of DNA (0.025%) codes for tRNA. The tRNA molecule consists of a single strand looped around itself with the 3' end always terminating in a C-C-A sequence. Some bases are paired.

The function of tRNA is to carry aminoacids to mRNA during protein synthesis, Each amino acid couples with a specific tRNA. Since the protein molecule consist of of about 20 types of amino acids, there must at least be 20 types or tRNA. But there could be many more, since there are two tRNAS for every amino acid in several, cases.

Mitochondria: (Gr. *mito* = thread, *chondiron* = granule)

Popularly called 'power house' of cells, these granular or filamentous organelles are present in all eukaryotes. They are characterised by their special structure, and unique physiological function concerning intracellular energy transformation.

Historically, mitochondria were first observed by Altmann (1894) who called them *bioblasts*. The present name was coined by Benda (1897) who developed the crystal violet staining. Bensely and Hoerr (1934) first isolated the liver cell mitochondria. The discovery that mitochondria are seats of cellular respiration was made by Hogeboom *et al* in 1948.

It is difficult to observe mitochondria in living cells due to their low refractive index, but they can be studied under phase contrast microscope after staining with a dilute solution of janus green. Mitochondria may be isolated from cells by centrifugation between 20,000 g -40.000 g.

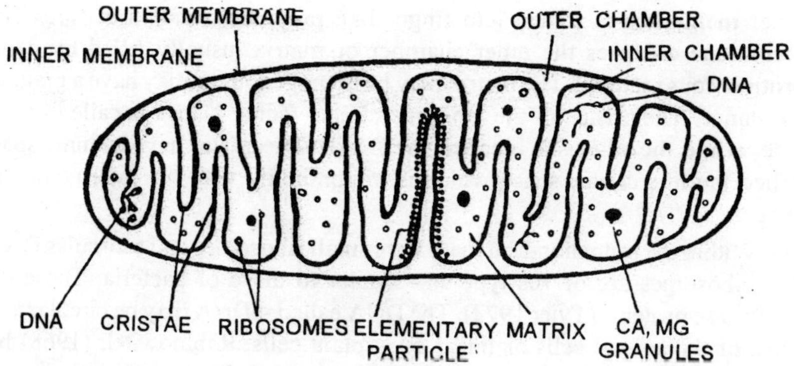

Fig. 2.24 The Cell

Diagrammatic representation of a section through a mitochondrion (ETPs are not shown on all christae)

Shape : Generally mitochondria are granular or filamentous, even through the shape is variable. During certain conditions a long mitochondrion may swell at one end lo become club shaped or look like a tennis racket .(De Robertis et al 1970). Mitochondria may also become vesicular.

Size : Size also is variable. The width is about 0.5μm and the length may reach a maximum of 7μm. The si/e depends on the osmotic pressure.

Distribution : In general, mitochondria are distributed evenly in the cytoplasm, but in some instances they may accumulate around the nucleus or at the periphery of the cytoplasm. During cell division, mitochondria accumulate near the spindles. The distribution of mitochondria within the cell is in relation to the energy demand. For example in skeletal muscles they are grouped like rings or braces around the I-band of the myofibril.

Orientation : Mitochondria have more or less a specific orientation. In cylindrical cells, their long axis is parallel to the long axis of lhe cell. In leucocytes, mitochondria are arranged radially in respect to the centrioles.

Number : Mitochondria are absent in prokaryotic cells. The quantity of mitochondria per cell varies, but in liver cells it is estimated that mitochondria contains 30-35% of the total protein content of the cell. A normal liver cell contains about 1000-1500 mitochondria. Some oocytes coniain an many as 300000 mitochondria.

Structure of mitochondria : The mitochondrion is enveloped by two membranes viz, *inner membrane* and *outer membrane.* The outer membrane is 6nm thick; internal to this is the inner membrane. Separating these two is a space (outer chamber) filled wilh a watery fluid. This space is 40-70Å thick. The inner membrane is folded into finger like projections called *cristae.* This membrane encloses the inner chamber or matrix, usually filled by dense protinaceous material. The matrix may be homogenous or may have a granular substance. The cristae of the inner membrane facing matrix is called the M side, while the other side is called the C side. The cristae have an inner space called the intercristae space which is in continuity with the intermembrane space.

Within the mitochondrial matrix are small ribosomes and a circular DNA. The ribosomes are of 70s type and similar to those of bacteria. These can synthesise proteins (Tyier 1973). The DNA called mDNA may be circular as in most of the animal cells or linear as in plant cells. Rabinowitch (1968) has demonstrated that the mitochondrial DNA has more G-C content than the nuclear DNA.

The mitochondrial DNA can replicate itself as it has DNA ploymerase. It is due to this, that mitochondria can replicate themselves.

The inner membrane : The inner membrane has numerous nail like or tennis racquet like projections protruding into the matrix. Each particle is about 70-100A^0 in diameter. The particles are placed apart a distance of 100A^0 and each is attached to the inner membrane by a short stalk,of 35-50A^0 long. The number of these particles per mitochondrion ranges from 10^4-10^5. These are called elementary particles, F_1 *particles* or *subunits*. It was formerly believed that these particles contain all the enzymes necessary for electron transport and oxidative phosphorylation. Hence they were named *electron transport panicles* (ETP). Each particle was thought to have four complexes, with complex I and II in base, III in stalk and IV in the head piece. According to current thinking, the head piece has ATPase, the stalk consists of a special protein (oligomycin sensitivity confering protein) and the piece contains the proton channel.

The inner membrane has the respiratory chain consisting of five complexes. Of these, four are involved in the electron transport, while the fifth one is involved in the synthesis of ATP.

Mitochondria play a vital role in respiration. Glycolysis takes place in the *cytosol* (cytoplasm) while the enzymes of the Kerbs cycle are found in the matrix, those of oxidative phosphorylation are found in the inner membrane.

Biogensis of mitochondria : Each mitochondrion lasts for about a week or little more. There are three views regarding the origin of mitochondria-a) origin from cell membranes, b) Division of pre existing mitochondria and c) *de novo* origin.

According to Robertson (1964), mitochondria in muscle cells arise by the infloding of the plasma membrane. This infolding joins another piece and forms a vesicular structure. In the nerve fibres of cray fish, endoplasmic reticulum may form mitochondria.

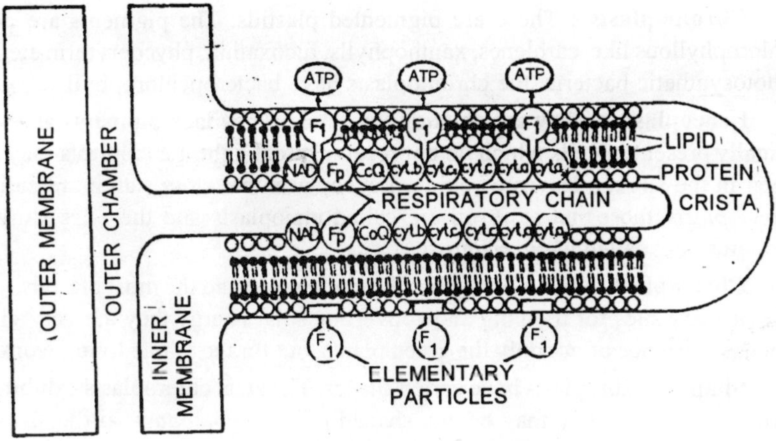

Fig. 2.25 The Cell

Arrangement of complexes (I-IV) in the ETP (old view)

According io Luck and Rich (1964), mitochondria are capable of self feproduction. DNA, protein synthesis, DNA polymerase and all the other essential, features for self repliciation are present in mitochondria.

Harvey (1951) supports a *de novo origin* for mitochondria. Centrifugal separation of sea-urchin eggs into two halves, a nucleated half (without mitochondria) and a non nucleated half (with mitochondira) showed that the nucleated half produced the mitochondria.

Are mitochondria symbolic organisms? The many similarities between a bacterial cell and mitochondria made many people (Lloyd 1974; Borst 1977, O'Brien 1977 believe that the mitochondrion was once an autonomous entity which later entered into a symbiolic association with an anaerobically respiring host cell. The host cell derived its energy only from glycolysis. Subsequently with the association, the host cell started respiring aerobically. De Robertis (1970) opines that in evolutionary terms it is possible that a symbiotic relationship could have evolved into the present semiautonomus condition of the mitochondria.

The same theory has been Proposed for chloroplasts also, as they also have all (the independent attributes of mitochondria.

PLASTIDS : These are living cytoplasmic inclusions found in most of the plants except for some bacteria and some fungi.

The plaslids are of three categories viz., chromoplasts, leucoplasts and chloroplasts.

Chromoplasts : These are pigmented plastids. The pigments are non chlorophyllous like, carolenes, xanthophylls, fucoxathin, phycoerythrin etc. In photosynthetic bacteria, the chromoplasts have bacteropchlorophyll.

Leucoplasts : They are colourless plastids. They lack pigments and are usually present in cells which do not receive direct light. Leucoplasts may be seen in the storage leaves of onion. Leucoplasts that store starch are called *amyloplasts,* those that store oil are called elaioplasts and the ones storing proteins are called *aleuroneplasts.*

Chloroplasts : These are by far, the commonest and the most plastids. As the primary sites for trapping and converting solar energy they are very vital for the existence of not only the green plants, but for the whole living world.

Shape : Chloroplasts have varied shapes. The algal chloroplasts exhibit an array of shapes. They may be cup shaped *(Chlamydomonas),* girdle shaped *(Utothrix),* reticulate *(Hydrodictyon),* stellate *(Zygnema),* ribbon shaped *(Spirogyra)* or discoid. They are generally spherical or ovoid in higher plants.

Size : There is a great variety in siz.e. Normally they are about 1μm thick and 4-6μm in length. Chloroplasts ofpolyploid cells are generally larger than in the diploid cells. Cells in the shaded area have longer chloroplasts but with less intensity in colouration.

Distribition : They arc uniformly distributed all over cytoplasm, but in some instances they cluster towards the nucleus. The concentration of chloroplasts will also depend on light intensity.

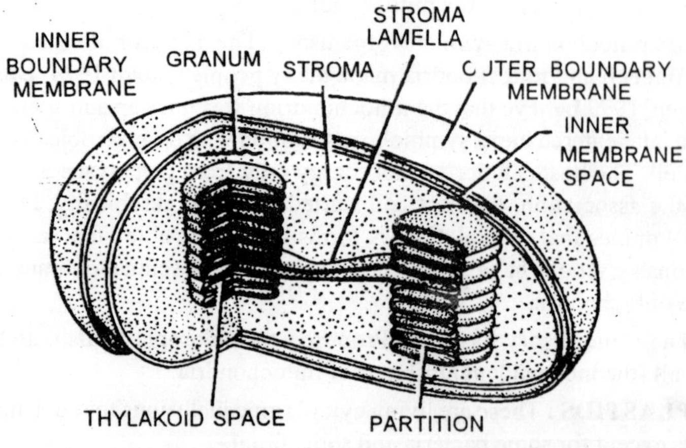

Fig. 2.26 The Cell
Model of Unltrastructure of chloroplasts

Number : The number of choloroplasts per cell varies. In some algae like *Chlamydomonas,* there is only one chloroplast per cell or two as in *Zygnema.* In higher plants, cells have a number of chloroplasts. The leaf of *Ricinus communis* has nearly lour lakh chloroplasts per square millimeter area.

Chemical composition : The chloroplasts contain carbohydrates, lipids, proteins chlorophyll, carotenoid and xanthophylls. The protein lipid ratio is 40:30. Besides the above, nucleic acids have also been demonstrated in the chloroplasts.

Ultra structure of the chloroplasts : The chloroplast has a covering of two membranes with an inner membrane space. These membranes are smooth and there are no perforations or particles. The membranes are differentially permeable (Mudrak).

A section of the chloroplast reveals an intricate system of membranes enclosed in a granular matrix. These membranes are called lamellae and the surrounding matrix-the *stroma.* In a sectional view, the lamellae can be seen packed and these stacks are called *thylakoids.* In the chloroplasts of higher plants, the thylakoids themselves form highly compact bundles called *grana.* Some thylakoids of granum extend into the stroma and maintain contact with other grana. These are *called-Stroma thylakoid* or *stroma lamellae* or *intergrana.*

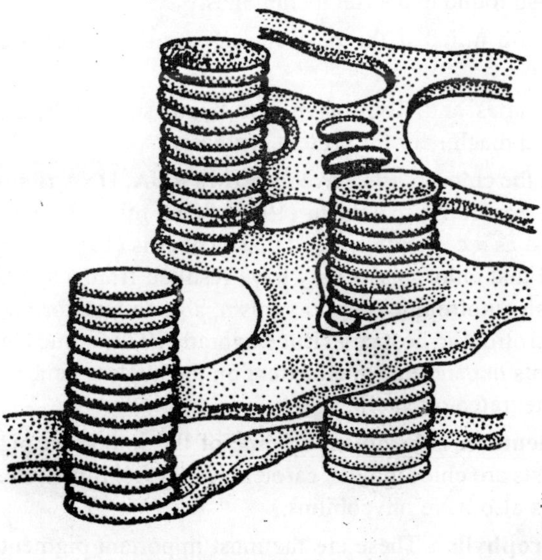

Fig. 2.27 The Cell
Three dimensional view of Grana (enlarged view)

The chloroplasts of algae lack the granum arrangement and in cyanobacteria (blue green algae), the thylakoids lie naked in cytoplasm without any envelope. In such instances, pigments are uniformly distributed on or in the lamellae.

The stroma has in addition to the lamellae - granules (globuli), lipid droplets, starch grains and vesicles.

Structure of the lamellae : All the photosynthetic pigments are concentrated in the chloroplast lamellae. The lamellae as well as the outer membranes are composed of lipoprotein sub units, each sub unit according to Weir and Benson has a protein core surrounded by a lipid sheath. The chloroplast membranes have a single layer of sub units, whereas in thylakoids it is double due to the juxtaposition of the membranes. The appressed areas are called partitions. These are hydrophobic in nature The space between the two membranes of thylakoid (fret membrane) is called *fret channel.* The space between two thylakoids is called *loculus* while the ends of disc shaped thylakoids are called *margins.* These areas (loculus, fret channel) are hydrophilic. The chlorophyll molecules are concentrated in the fret channel ie., in the space between the two membranes of thylakoids. The chlorophyll molecules may be entirely in the space or the *heads* may be in the fret channel while the *tails* are buried in the sub units, occasionally the entire chlorophyll molecule may be found in the fret membrane. Other pigments like cytochromes, carotenoids, etc., are also found in the fret membranes.

Four sub units of the partition constitute a photosynthetic unit or a *quantosome.*

Ribosomes and RNA have also been isolated from the chloroplasts indicating a machinery for protein synthesis.

Since the chloroplasts, have their own RNA, DNA ribosomes and protein synthesis, according to Deviln (1975) "One might be inclined to view the chloroplast as a cell within a cell". De Robertis (1980) suggests "It has been suggested that chloroplasts may have resulted from a symbiotic relationship between an autotrophic microorganism, able to tmasform energy from light, and a heterotrophic host cell" But Bogorad (1977) opines that the chloroplast components depend to a great extent on the nuclear genes for their assembly into an integrated organelle.

Pigments in chloroplast : Some of the important pigments present in chloroplasts are chlorophylls, carotenoids, cytochromes etc. In algal cells the thylakoids also have phycobilins.

Chlorophylls : These are the most important pigments involved in the basic photochemical reaction that sustains life. Atleast nine types of chlorophylls have been identified so far - chlorophylls *a. b. c. d* and *e.* Bacterio Chlorophylls a and b and the chlorobium chlorophylls 650 and 660μm.

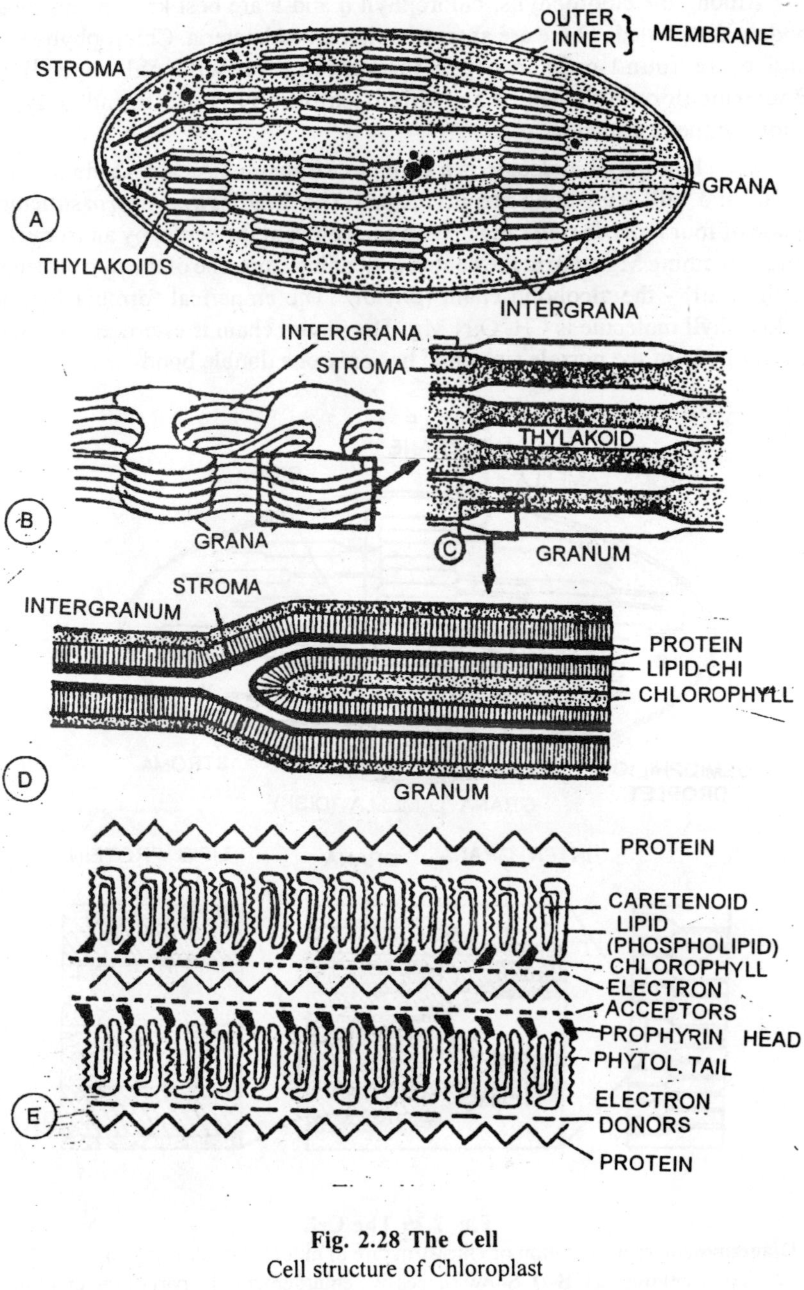

Fig. 2.28 The Cell
Cell structure of Chloroplast

Among the chlorophylls, chlorophyll *a* and *b* are best known and most widely distributed. These are absent in pigmented bacteria. Chlorophylls *c, d* and e are found only in algae in combination with chlorophyll a. Bacteriochlorophylls and chlorobium chlorophylls are found only in photosynthetic bacteria.

The chlorophyll molecule is a complex structure basically made up of a 'head' and 'tail' resembling a tennis racquet. The 'head' is a *porphyrin* structure made of four *Pyrrole rings* attached to each other at the centre by an isocyclic ring containing Mg atom at the centre. Extending from one of the pyrrole rings is the 'tail' - the alcoholic chain *(phytol).* The emperical formula for the chlorophyll molecule is CH, O,N Mg. The phytol chain is esterified on the C atom of one of the pyrrole rings and has only one double bond.

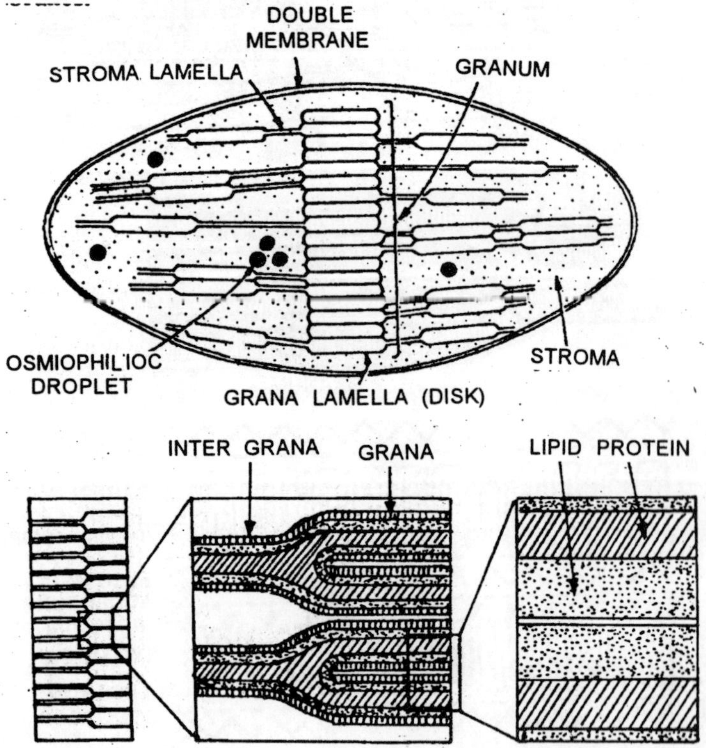

Fig. 2.29 The Cell

Diagrammatic representation of Ultrastructure of chloroplast showing the lamellae. **A.** Entire chloroplast, **B-D.** Show successive enlargement of a part of the granum

Chlorophyll *a* and *b* differ structurally in having different atomic groupings at the C atom in the pyrrole ring. In chlorophyll a C atom has a methyl group, while chlorophyll *b* has an aldehyde group; besides the two pigments have different absorption spectra. The peaks are as follows

Chlorophyll *a* 429, 410 and 660nm

Chlorophyll *b* 430,453 and 442nm

(The above spectra are for chlorophyll in *vitro)*

Carotenoid pigments : These are lipid compounds ranging in colour from yellow to purple. They are widely distributed in both plants and animals. They are also present in microorganisms including red algae, cyanobacteria, photosynthetic bacteria, fungi, etc. Wackenroder (1931) isolated the first carotenoid - carotene from tissue.

Carotenoids are derivatives of *lycopene* a red pigment found in many plants. They (carotenoids) are known to have eight isoprene (CH_2 = $C[CH_3]$ - CH = CH_2) like residues.

The main carotenoid found in plants is the orange yellow coloured β carotene with some quantity of α carotene.

The carotenoids are also located in the chloroplast. According to Goodwin (1960), carotenoids and chlorophylls may be combined with the same protein to form a complex known as *photosynthein.*

Carotenoids protect the chorophyll form photooxidation and transfer the light energy they absorb to chlorophyll *a.*

Phycobilins: These are found only in algae. There are two types of pycobilins - the *Phycoreythrin* (red) and *Phycocyanin* (blue) These pigments are strongly associated with a protein and consequently it is difficult to isolate the pigments in the pure state.

Like carotenoids phycobilins are also involved in the transfer of light energy (they absorb) to the chlorophyll.

The absorption spectra of phycobilins presents an intersting study considering the fact that they act as accessory pigments in photosynthesis. R - phycoerythrin has peaks at 495, 540 and 545nm, while R - phycocyanin has peaks at 550 and 615nm.

ENDOPLASMIC RETICULUM : The cytoplasm in the cells of most of the plants and animals is traversed by a network of anastamosing fibrillar structures called endoplasmic reticulum (ER). Originally known as ergastoplasm, the current name ER was given by Porter and Kaallman (1952). The ER is connected to the outer nuclear membrane with the ER space opening into the perinuclear space between two membranes. The ER is known to be connected with the plasma membrane.

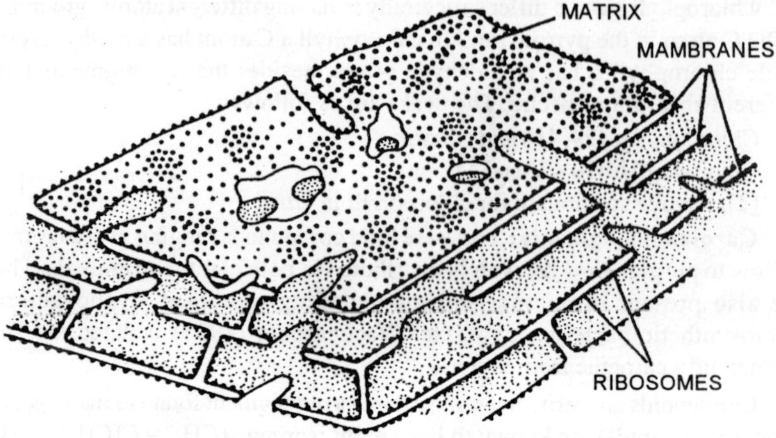

Fig. 2.30 The Cell
Three dimentional view of endoplasmic reticulum

Morphologically the ER consists of three types of structures viz., *Cistemae, Vesicles* and *Tublllues.*

The **Cisternae** *are* elongated cyllindrical unbranched structures arranged parallel to one another respresenting a lamellated structure. They are 40 - 50μm in diameter. Cisternae are generally found in cells active in synthesis.

The **Vesicles** or sacs or oval bodies, attached to the membrane and have a diameter of 25-500μm. These are abundant in pancreatic cells.

The **Tubulues** are branched irregularly and are of diverse shapes having a size of 50 - 100μm. The tubules are abundant in cells which are active in the synthesis of steroid compounds. They are also found in the epithelical cells of retina.

Ultra structure : All the three components of ER are bounded by a unit membrane 50 - 60Å thick. The membrane is trilamenar like the other unit membranes. Palade (1956) has observed some secretory granules in the cavity of endoplasmic reticulum. The membrane has two outer layers of protein enclosing two layers ofphospholipids. However the ER membrane is symmetrical unlike in other organelles (mitochondria).

Types of ER : Two types of ER have been observed in the cells (same or different). These are - smooth ER (agranular) and rough ER (granullar).

The membrane of smooth ER are not granular due to the absence of ribosomes. Smooth ER in mainly seen in tubules. The tubules measure about 500 - 1000Å in length. Agranular ER is seen in cells involved in lipid or steroid synthesis. Cells active in protein synthesis do not have agranular ER. Agranular reticulum of muscle cells is called *sarcoplasmic reticulum* and it forms lace like sleeve around myofibrils. In the pigmented cells of retina, the agranular ER occurs in the form of tightly packed tubules and vesicles called *myeloid bodies*.

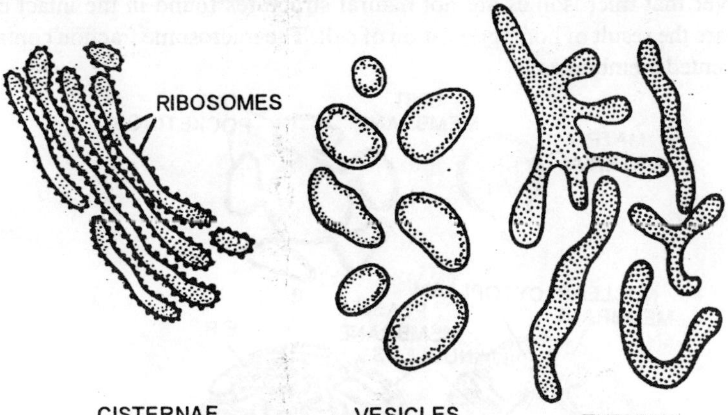

Fig. 2.31 The Cell
Various components of endoplasmic reticulum

The rough ER has a granular surface due to the presence of a large number of ribosomes attached to it. Since ribosomes are actively involved in protein synthesis rough ER is conspicuous in cells that are actively involved in protein synthesis e.g. meristematic cells, plasma cells, goblet cells etc. The rough ER stains positively by basic dyes.

The cells (Spermatocytes) of certain vertebrates show an unnsual type of ER. Usually ER membrane does not have pores or openings. But in the instance mentioned above pores or annuli have been reported resembling the ones found in the nuclear membrane. There is a diaphragm across the octogonal symmetry (Manul 1968).

The ER performs many vital cellular functions. These are

(a) Enzymatic activities
(b) Transport of synthetic products
(c) Storage of metabolities
(d) Formation of nuclear envelope
(e) Intracellular impulse conduction etc.

Microsomes : These are structures that result due to the homogenization of the membranes. The cell membranes break up into fragments, round off and form microsomes. Microsomes are actually fragments of ER. They can be isolated by high speed centrifugation (100,000g) when they settle after nuclear and ribosomal fractions. Observed under electron microcope they appear as membrane bound vesicles of 500- 1500 Å in diameter. It should be understood however that microsomes are not natural structures found in the intact cell. They are the result of homogenization of cell. The microsome fraction contains fragmented membranes.

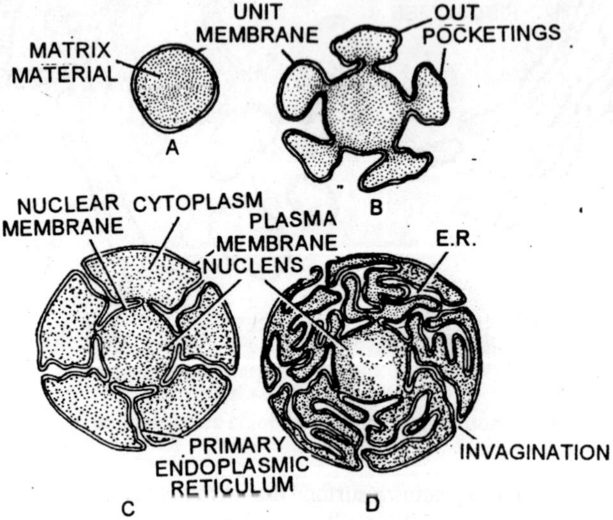

Fig. 2.32 The Cell
Stages in the origin of Endoplasmic reticulum

Origin of ER : The origin is uncertain, invagination of the plasma membrane might give rise to a canalicular system. But evidences for this hypothesis are yet to come (Dallmes et al, 1966).

RIBOSOMES : These are dense granular nuclecoprotein structures occuring in cytoplasm, matrix of mitochondria and chloroplasts. In many instances ribosomes are attached to the ER. Observed first in plant cells in 1953 by Robinson and Brown while studying bean roots, these were later discovered in animal cells also. Ranging in diameter from 150 - 200Å they have RNA and protein in equal quantities.

Ribosomes are isolated by differential centrifugation depending on sedimentation coefficient. The sedimentation coefficient is expressed in terms of *svedberg units.* The S units are related with the size and weight of the ribosome molecules.

Types of ribosomes : Two types of ribosomes have been identified based on the sedimentation coefficient which shows how soon a cell organelle can sediment under centrifugation. If the organelle is heavier, its sedimentation coefficient is more. The two types of ribosomes are - *70s ribosomes* and *80s ribosomes*.

The 70s ribosomes are comparatively smaller and have a molecular weight of 2.7 x 10^6 daltons (Dalton is a unit of molecular weight; one dalton = weight of one H atom = 1) According to Huxley and Zubay (1960) the size of a 70s particle is 170 x 170 x 200Å. These ribosomes occur in prokaryotic cells as well as in chloroplasts of eukaryotic cells. The 70s ribosomes have two sub units - a large 50s sub unit and a small 30s sub unit.

The 80s ribosomes have a moleculer weight of 4 x 10^6 daltons, and are predominantly found in eukaryotic cells. There are two sub units in an 80s ribosome. These are - 60 and 40s.

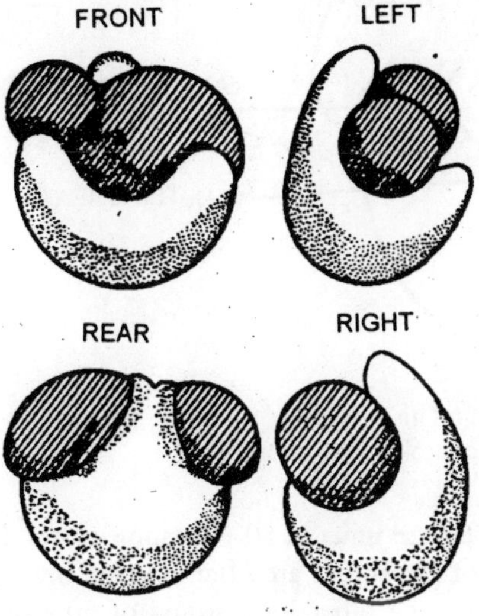

Fig. 2.33 The Cell
Ribosome - different views

Ribosomes may occur singly as isolated units when they are called *monosomes*. When they occur in clusters or groups, they are called *polyribosomes. Tbe* polyribosomes may have a sedimentation coefficient of 100s - 600s.

The number of ribosomes per cell varies, it may be 10,000 (bacterial cell) or up to 10 million (eukaryotic cell).

Ribosomes of chloroplasts and mitochondria have their own protein synthesis. They have a sedimentation coefficient of 55s with two sub units 40s and 30s.

Ultra structure of ribosomes : Ribosomes are oblate or spheroidal structures having two sub units (a large and a small). The large sub unit is dome like and the smaller subunit is placed above like a cap. The 70s ribosome has two units 50s and 30s as already mentioned. The 50s sub unit has a diameter of 140 -

160A°and is pentagonal in shape. In the center of this sub unit occurs a round area of 40 - 60A°. According to Fernando (1968), there is a pore like transparent area in the centre which prohibits the entry of proteolytic enzymes.

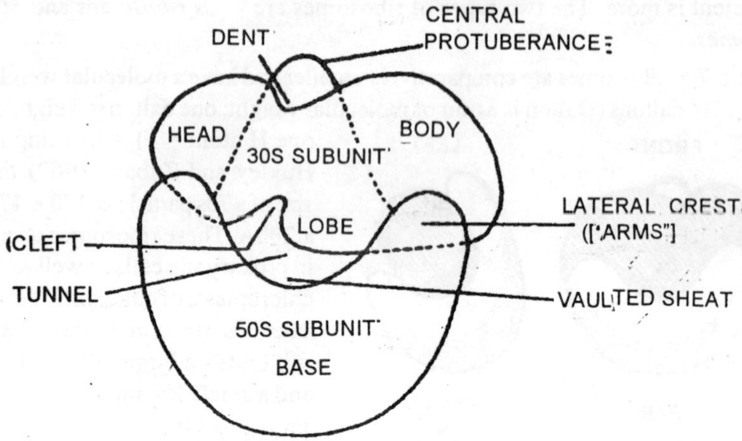

Fig. 2.34 The Cell

Witlmann's model of prokaryotic ribosome

The pores have been found in 50s sub units of 80 ribosomes also. The smaller sub units of both the types of ribosomes are often divided into two portions which remain connected to each other by a strand of 30 - 69A° Thickness.

Models for ribosomes structure : There are two models, viz., Wittmann's quasi symmetrical model and Lake's asymmertrical model.

According to Wittmann's model, the 30s sub unit is an elongate slightly bent stucture, having a head and a larger body with a cleft in between; with two parts protruding uniequally. It appears like a telephone receiver. The 50s sub unit is variously shaped - round shaped, maple leaf .shaped etc. The 50s sub unit has three protruberances.

The two sub units of 70s ribosomes have four areas of contact. The mRNA binding. region is located on the right side of the *30s* sub unit.

Lake's model regards the 30s sub unit to be totally asymmetrical. The sub unit is divided into two parts (1/3 and 2/3) unequally. Extending from the lower 2/3 region is a portion called the *platform.* This cleft is important in the sense of location of codon - anticodon interaction. The major difference between the Wittmanns model and Lake's model for the 30s sub unit is that the latter regards the 30s sub unit as totally asymmertrical. According to Lake, even the 50s subunit is asymmetrical.

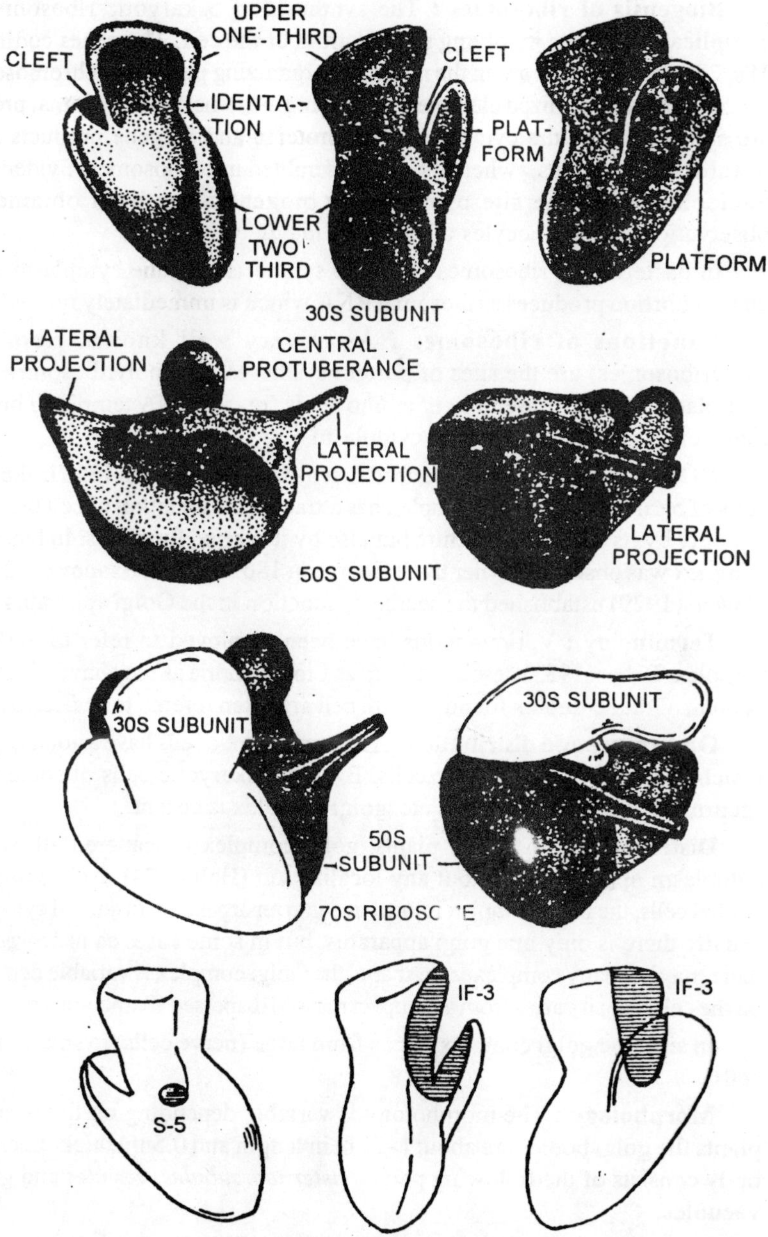

Fig. 2.35 The Cell
Lake's model of 70s ribozome from a prokaryotic cell

Biogensis of ribosomes : The synthesis of eukaryotic ribosome is a complicated process involving several areas of the cell. The genes coding for 18s, 28s and 58s RNA are in the nucleolar organizing part of the chromosomes; the 5s genes are localized eleswhere in the chromosome. The ribosomal proteins are synthesised in the cytoplasm. The proteins and 5s gene products move towards the nucleolus, where they are assembled into ribosomes. Evidence for nucleolus being the site of ribosome biogeneis has been obtained by observations on the oocytes of *Xenopus laevis.*

In bacteria, the ribosomes (70s) are synthesised in the cytoplasm as the cistron portion produces a ribosomal RNA which is immediately relased.

Functions of ribosomes : As is very well known, ribosomes (polyribosomes) are the sites of portein systheses. The polyribosomes serve as a platform for the asembly *of* amino acids (on a mRNA template) brought together by specific tRNA from cytoplasm.

GOLGI COMPLEX : Described first by Camillo Golgi in 1890 in the nerve cells of barn owl, the Golgi complex has attracted the attention of cell biologists not only by its complex structure but also by its varied functions. In fact Golgi complex was observed earlier by St George in 1865 itself. Nassonov (1923) and Bowen (1929) established the secretory function in the Golgi apparatus.

Terminoloy : Various terms have been employed to refer to the Golgi complex. Baker (1953) described them as Lipochondria as they have a high lipid content. Golgi complex found in plant cell are often refered to as *Dictyosomes.*

Occurrence and distribution : All the eukaryotic cells have a golgi complex which is absent in prokaryotic cells. Even in eukaryotic cells of some fungi, spenrmatozoids of bryophytes etc. golgi complex is absent.

Distribution : In higher plants, golgi complex is scattered all over the cytoplasm apparently without any localization (Hall 1974). As a contrast in animal cells, the golgi complex is localized into an organelle. In animal cytoplasm, usually there is only one golgi apparatus, but in some cases as in *Storeomyxa* there may be many complexes. In shape, the Golgi complex is variable depending on the cell type; it varies from a compact mass to dispersed filamentous reticulum.

In size, the golgi complex varies from large (nerve cells) to small (muscle cells).

Morphology : The morphology is variable depending on the source. In plants the golgi bodies are about I - 3nm in length and 0.5nm thick. Each golgi body consists of the following parts - *cisternae, tubules, vesicles* and golgian vacuoles.

The *cisternae* are flattened sacs filled with a fluid. Each golgi body consists of 4 - 8 flat structures which are slightly curved and are arranged in stacks. Each cisterna is enclosed by the trilaminar membrane 6nm in thickness and is separated from the other by a 100 - 150Å wide space. In some instances the intercisternal space is filled by fibres called intercistemal elements. The membranes of the cisternae are *fenestrated* (porous). The pores may be localized or present all over the cistemae. Occasionally cisternae may form reticulate structures.

Tubules: Arising from the peripheral region of the cisterane, the tubules form an anastomosing network of 300 - 500Å in diameter.

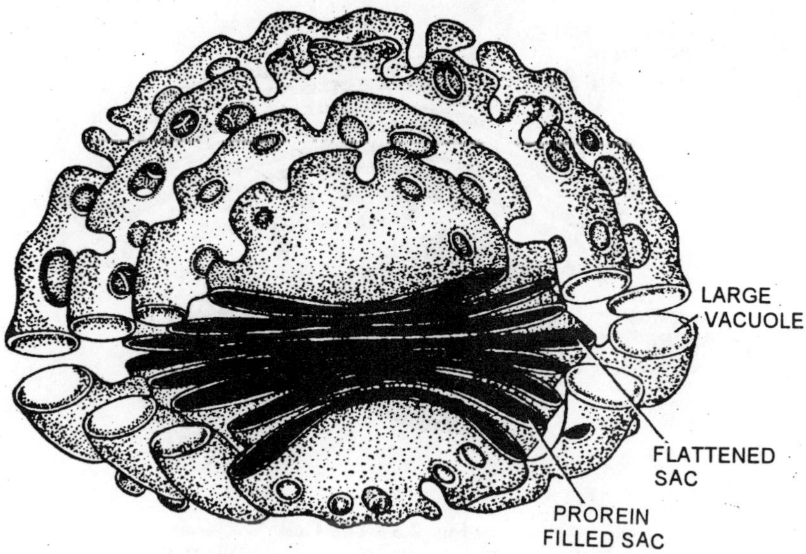

Fig. 2.36 The Cell
Golgi Complex

Vesicles : These are goblet like structures attached to the tubules. There are two types of vesicles *smooth vesicles* and *coated vesicles.* The smooth vesicles are 20 - 80nm in diameter and are often called secretory vesicles. These are pinched off from the ends of cistetnal tubes. The coated vesicles are spherical outgrowths about 50nm in diameter. The function of the coated vesicles is not known.

Vacuoles : These are large spaces occuring at the distal end of the cisternae. They represent the cisternae whose membranes are widely separated.

In the case of Dictyosomes (isolated golgi bodies) of plants, cisternae are extensively fesnestrated and have rough and smooth vesicles attached to them. Parallel fibres fill the intercisternal place.

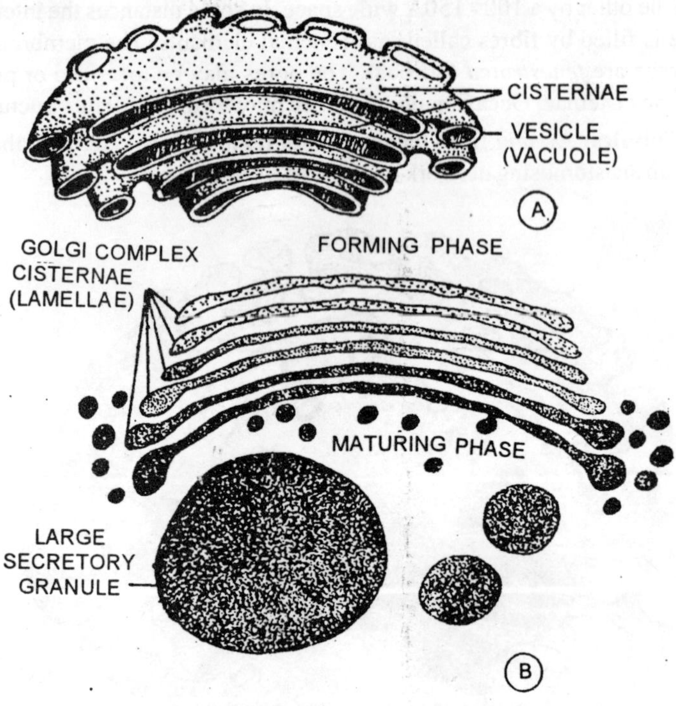

Fig. 2.37 The Cell
Golgi Complex (another view) **A.** Steroscopic view, **B.** Section view

Chemical composition of golgi complex : Proteins and phospholipids are in equal concentration. Golgi material in the nerve cells consists of cephalin and lecithin. Giroud has reported Vitamin C in the golgi complex.

Functions of the golgi complex : The following are some of the functions attributed to the golgi complex -

1. Absorption of compounds
2. Sites of enzyme production
3. Sites of hormonal production
4. Sites of protein storage
5. Forms acrosome during sperm maturation
6. Forms intracellular crystals
7. Formation of plant cell wall - the golgi bodies synthesise pectin, hemiceullulose and cellulose microfibrils. They also help in the formation of cell plate during mitosis.

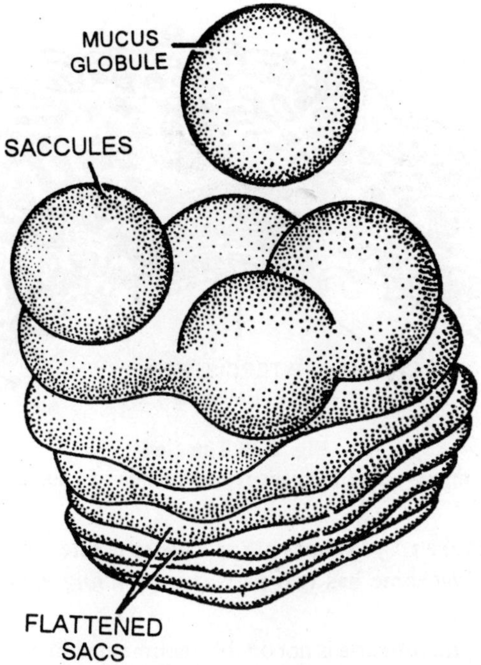

Fig. 2.38 The Cell
Golgi Complex with mucucus globules

Origin of golgi complex : Evidences indicate that the golgi bodies are produced from the membranes of the smooth ER. Earlier it was thought that they arise *de novo* or from the phragmoplasts of the cell wall. Both these theories are found to be incorrect.

Lysosomes : (*Gr.Lyso* = digestive, *soma* = body). First discovered in 1949 as dense round bodies, the lysosomes were called *pericanalicular bodies*. It was de Duve, (1955) who gave the present name.

Occurence : Lysosomes are found in most of the animal cells and few plant cells. The hepatic cells, pancreatic cells, kidney cells etc. posses large number of lysosomes. According to Matile (1969) the lysosomes in plant cell are membrane bound. Generally lysosomes are absent in the prokaryotic cells. Slime molds also have lysosomes.

Lysosomes are pleomorphic as seen in the meristematic cells of the root. The diameter may be 0.4 to 0.8μ; but in some mammalian kidney cells lysosomes as long as 5μ have been found.

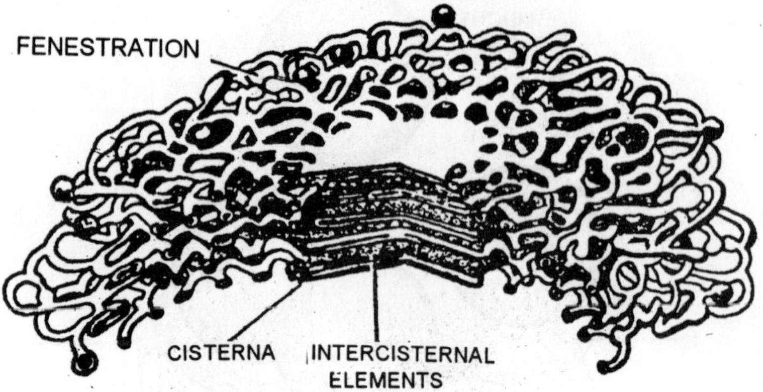

Fig. 2.39 The Cell
Golgi Complex – Diagrammatic representation of a plant dictyosome

Ultra structure : Lysosomes are saccate structures filled with digestive enzymes. Each lysosome has two parts, viz., *limiting membrane* and *inner dense mass.*

The **limiting membrane** is not double membraned like in other organelles. It is single and composed of lipoproteins. The membrane however is bilayered like any unit membrane.

The **inner dense mass** is the core of lysosomes; it is mostly enzymatic. The membrane is impermeable to substrates the enzymes of which are present in the lysosomes. Occassionally certain substances called *labilizers* destabilize the membrane causing the enzymes to come out. Alternatively, there are substance called *stabilizers* which stabilize the membrane. The limited permeability of the membrane prevents excessive digestion of cellular components by the lytic enzymes of lysosomes. The enzymes present in the lysosomes are - Nucleases, Phosphatases, Lipases, Glycosidases, and Sulphatases.

Polymorphism : Lysosomes are ploymorphic assuming various shapes depending on the cell type and situation. Basically there are two types of lysosomes viz. *Primary lysosomes and secondary lysosomes.* The secondary lysosomes are of three categories - *Phagosomes, residual bodies* and *autophagic vaculoes.*

Primary lysosomes : There are sac like structures produced directly from ER from the Glogi bodies.

Phagosomes : These are also known as heterophagosomes or digestive vaculoes. When a foreign substance enters the cytoplasm, the plasma membrane forms a sac around these substances. These are the phagosomes or pinosomes.

The lytic enzymes present in the phagosome digests the foreign substance. The phenomenon is called *Phagocytosis* or *pinocytosis*.

Residual bodies : After the digested material has passed out of the lysosomes, the undigested material remains within the lysososmes. Such lysosomes with residual substrate are called residual bodies. Residual bodies may remain within the cell or very often they are pushed out of the cell by exocytosis.

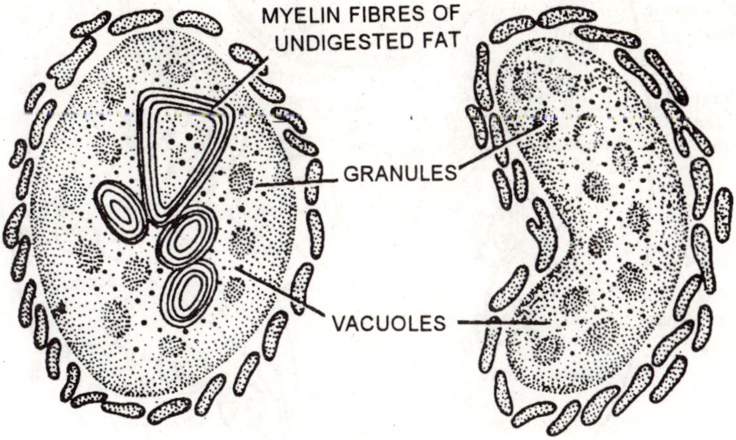

Fig. 2.40 The Cell
Lysosome – Structure

Autophagic vaculoe : Occasionally, a lysosome may digest a part of the cell (ER, mitochondria, etc.). This is a mechanism to digest a part of the cell without destroying the entire cell. Liver cells show autophagic vacuoles or autophagosomes under starvation.

Origin of lysosomes

(a) *Extra cellular origin* : Vacuoles absorbed due to'pinocytosis or phagocytosis may become secondary lysosomes after fusing with primary lysosomes.

(b) *Origin from golgi bodies* : There are enough evidences to show that lysosomes are produced from golgi bodies at the height of their secretory activity.

(c) *Origin from ER* : According to Novikoff (1965), the rough ER gives rise to the lysosomes.

Major functions of lysosomes : The main function of the lysosomes is digestion of intracellular substances. According to Hirsch and Coh (1964), lysosomes help in destroying the dead cells from the tiusse. Recently it has been found out that in some animals (frog), lysosomes play an important role in metamorphosis. Lysosomes also perhaps help in the formation and destruction of bone cells.

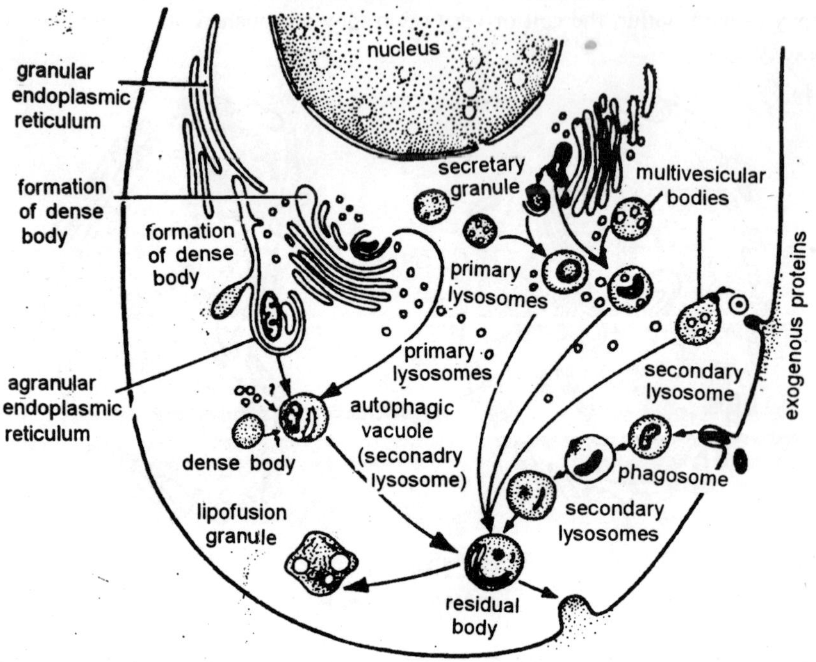

Fig. 2.41 The Cell
Polymorphism in Lysosomes

CENTROSOMES : The centromes are found in the centre of the cell near the nucleus. These are present in all animal cells and lower plants like Dinoflagellates, *Euglena* and *Chlamydomonas,* etc. Van Benden (1887) disocvered centrosomes.

The centrosomes play an important role in organising the mitotic apparatus in flagellated animals, while in flagellated plants they function as basal granules of flagella.

Size : In mature centrosomes the diameter is 1600Å - 2500Å while the length is about 160A^{0}. Giant centrioles have been found in spermatocytes or oocytes of certain organisms.

Origin: They have a multiple origin.

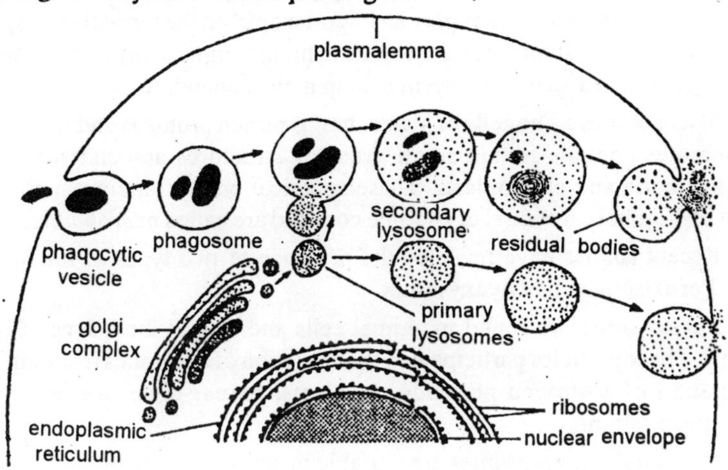

Fig. 2.42 The Cell
Origin and phases of Lysosomes

The centrosome has two parts, viz., *centrioles* and *centrosphere*

Centrioles : These range in number from one to two and are made up of RNA. Electron microsocopic observation has revealed the centrioles to be paired cylindrical structures 3000 - 5000Å long and 1200 - 1500Å in diameter. These are open at one or both ends and placed at right angles to each other. Each cylinder is composed *of* nine fibres spaced equally and running parallel to the cylinder axis. Each fibre is in turn made up of three *microtubules (triplet).* In a cross section, these appear to be arranged like the vanes of a pin wheel. The vanes are tiled at an angle of 40^0. The wall of the cylinder is constituted by these nine triplet fibres.

The microtubule triplet in a fibre can be designated as A,B and C of which A is innermost. Each microtubule wall is $50A^0$ in diameter and the lumen in 120 - $130A^0$ (the total diametre varies from 180 - $200A^0$). The length of each microtubule is about $560A^0$. At the innerside of each triplet fibre a dense line runs and it is called *triplet base* connecting A and C tubules. In addition to this, a dense RNA region called *foot* is present.

Centrosphere : The dense granlar region of cytoplasm surrounding the centrioles is called the centrosphere.

Structure : Structurally the microtubules are lipoproteinaceous (Fulton 1971).

Functions : Centrosomes appear to play an important part in cell division. people believe that centrioles play an important part in the formation of spindle. They however constitute the mitotic pole in higher animals. In primitive animals which are ciliated, centrioles seem to help in the generation of cilia.

Microbodies : The cells of plants, fungi, certain protozoa and liver cells of vertebrates contain certain small granular structures associated with ER, mitochondria and chloroplasts. These spherical bodies ranging in size from 0.2μ to 1.5μ have a central crystalline core and are called *microbodies.*

Recent studies have revealed the presence of two types of microbodies viz., **Peroxisomes and Glyoxysomes**.

Peroxisomes are found in animal cells and also in the leaves of higher plants. These particles participate in the oxidation of substrates resulting in the formation of hydrogen peroxide. In plants, peroxisomes are involved in photorespiration.

Structurally peroxisomes are variable in shape and size but are generally circular. They have an envelope of a single membrane made up of proteins and lipids enclosing a granular matrix. In certain instances, the matrix is crystalline or amorphous. The peroxisomes catalyze the H_2O_2 production. In plants, the peroxisomes contain enzymes of glycolate pathway.

Glyoxosomes are found in the cells of yeast and certain other fungi *(Neurospora)* and also seeds (oil rich) of many higher plants. They are similar to peroxisomes except that their core çonsists of dense rods of 6.0μm in diameter.

Functionally the glyoxysomes contain enzymes for fatty acid metabolism, involved in the conversion of lipids to carbohydrates during germination. The *glyoxysomes* play an important role in lhe *glyoxylate cycle.* The enzymes *Urate oxidase* and *allantoinase* are also found in glyoxysomes. Urate oxidase converts uric acid to allantonin, while allantoinase hydrates allantonin to allantonic acid during the degrading of purines.

3

SOLUTIONS SUSPENSIONS AND COLLOIDS

All the physiological reactions in the cell of an organism take place in a watery medium. Water however is not present in the pure state in the cell. It is present either in the form of solution, suspension or colloids. Since these three forms of aqueous system form the very basis of all types of biological reactions, an attempt is made in this chapter to understand the basic characters of these three forms of liquid.

SOLUTIONS

Adding of any solid substance into a liquid results in a solution. For example, if a spoonful of salt or sugar is added to a glass of water, and stirred, in a little time the sugar disappears (dissolves) into water and a sugar solution is formed. In this solution there are two components - a substance which is dissolved and a medium in which the substance dissolves. The substance is usually a solid and the medium is a liquid, The solid substance is called the solute and the liquid is called *solvent.* The resultant product is the solution. In a solution, the solute and solvent particles are distributed uniformly resulting in a homogenous mixture. The solute and solvent molecules are constantly in motion, though irregularly.

Depending on the amount of solute present in the solution, it may be classified into dilute, concentrate and saturated. In a dilute solution only a little quantity of the solute is present. The solution becomes concentrated with increase in the amount of solute (with reference to solvent). In a saturated solution, the solute has reached its maximum quantity and no more dissolution of the solute is possible in the solution.

The solutes (solids) that dissolve in a solvent to form a soltution are of two kinds, *ionic* and non *ionic.* Sugar (sucrose) is a non ionic substance and hence its molecules do not undergo any disassociation but just get dispersed in the solvent. The common salt (sodium chloride) on the other hand is an ionic compound, whose molecules get ionized in water to sodium ions and chloride ions.

Types of solutions

Solutions are classified into three categories based on the size of the particle that gets into a solvent. These are

(a) Suspensions : The solute particles are greater in size than 0.1 μm. The solute does not separate into its molecules, hence it is not completely miscible in the solvent. The particles of the solute can be seen with naked eye and are dispersed throughout the system. A suspension is an unstable mixture. After some time, the solute particles settle down in a place, with the result, we can separate a pure solvent easily from a suspension just by decanting.

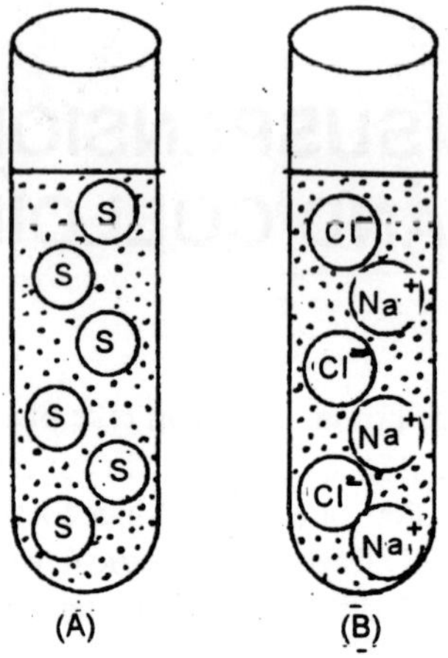

Fig. 3.1 Solutions, Suspensions and Colloids
Non ionic (A) and (B) Solutions

(b) Colloidal solutions : The size of the solute particle varies in size from 0.001μm to 0.1μm in diametre. The solute particles break up into molecules in a colloid but the molecules aggregate in small colonies.

(c) True solutions : In true solutions, the solute particles will be less than 1μm in diametre. They are distributed homogenously throughout thc solution. The system is stable in that the particles do not settle down in any one place. The solute particles will be either in molecular (non ionic) form or ionic form.

Concentration of solutions : Concentration (of a solution) means, the relative proportion of solute and solvent in a solution. The concentration of a solution may be expressed in any one of the following ways.

(i) Percentage of a solute by weight or by volume. (percentage method)
(ii) Molar solutions
(iii) Molal solutions
(iv) Normal solutions.

Percentage solutions : In a percentage solution, a particular weight of the solute is added to a particular quantity of solvent. For instance 20% solution means 20 gms of a solute (salt, sugar etc) in 80 gms of water. This is 20% solution W/W (weight by weight). In 20% solution weight by volume (W/V) 20 grams of solute is dissolved in 100 ml of solvent. There is another kind of percentage solution (used for liquid in liquid) wherein the measurement is volume by volume (V/V). For example 80% alchol means 80 ml. of alcohol in 100 ml. of solution.

Molar solutions : Molarity of a solution means, the gram molecular weight of a solute dissolved in a solvent to *make* the total volume to 1 litre. Gram molecular weight of a substance means weight in grams of that substance equivalent to its atomic weight in units. For example if the molecule of a substance contains six carbon atoms, 12 hydrogen atoms and six oxygen atoms; the total atomic weight would be 180.16; and the gram molecular weigh is 180.16 grams.

If 180.16 grams of Glucose ($C_6H_{12}O_6$) is dissolved in water to make the final volume 1 litre, it will give a solution which has I mole of glucose in 1 litre of water. Hence the solution is called 1 molar solution (1M). If 360.32 gms of glucose is dissolved in 1 litre of water it would give 2 M solution.

Another significant aspect of molar solutions is that equal volume of different solutions of same molarity will have the same number of solute molecules, but different number of solvent molecules.

Molal solutions : This is slightly different from molar solution in that, a gram molecular of weight of a substance is dissolved in I litre of a solvent (In a molar solution, gram molecular weight of a substance is dissolved to makeup the volume to 1 litre) so that final volume is more than 1 litre. The excess of solvent (more than 1 litre) depends on the solute used. Equal volume of molal solutions do not contain the same number of solute molecules.

Normal solutions : A normal solution is one, in which a gram equivalent weight of a substance is dissolved in one litre of solution. One gram equivalent gives a strength of 1N, 2 gram equivalent weight 2N solution etc. What is gram equivalent weight? The weight in grams of the element that combines with or equivalent to 1.008 grams of hydrogen. A normal solution may be obtained by dissolving the *molecular weight* quantity of the substance to make one litre of the solution. For example a 1N solution of NaOH can be prepared by dissolving 40 grams of NaOH in water and making the volume up to 1 litre

Molecular weight of NaOH = 23 + 17 = 40

Hydrogen equivalent = 1

Acids, Bases and salts : A plant, during its metabolic activities produces a great variety of substances which may be categorized either as acids or bases. Among the acids, the most important ones are Amino adds, carboxylic aicds. Fatty adds etc. Among the bases, the nitrogen bases namely purines and pyrimidines are most important. Acids are sour in taste and can change the colour of natural dyes from blue to red. Bases have a bitter taste and feel soapy in a solution.

A salt is formed when a base reacts with an acid. Salt and water are formed as a result.

Chemical nature : An acid may be defined as a molecule or ion that can donate a proton (H^+) to any other molecule or ion. When an acid is dissolved in water it undergoes *Ionization.* Ionization may be defined as the dissociation of a molecule into negatively and positively charged partides called *ions.* The negatively charged ions are called *anions,* and the positively charged ones are called *cations.*

A base may be defined as a molecule that will accept a proton.

Electrolytes : Certain compounds, when dissolved in water can conduct electricity and consequently undergo ionization. Such compounds are termed *electrolytes,* and the molecular dissociation is called *electrolysis.* Acids and bases are electrolytes. Sugars, alcohols and certain other substances are non electrolytes.

Amphoteric compounds : Compounds which act as both acid and base are called ampboteric compounds. For example water it is both a weak base and a weak acid. Amino adds are also amphoteric compounds.

Neutralization : When an equivalent amount of acid and base are mixed, acidic and basic properties are lost and *neutralization* is said to have taken place. This is due to the free hydrogen ions of the acid reacting with the hydroxyl ions of base to form water.

Hydrogen ion concentration : (pH)

The hydrogen ion concentration of any solution is called the pH. It wasArrhenius, who advanced the idea ofelectrolytic dissociation of acids and bases in water to form ions. But the whole of acid or base will not be present in ionic form. Some quantity is still present in molecular form. For instance, a normal solution of HCI has only 78%- molecules in ionized condition. The term normal is employed to the whole of hydrogen, whether present in the form of ions or in combined form. Hence in terms of hydrogen ions, a solution of HCI is only-0.78 N.

All aqueous solutions and water normally contain some concentration of H^+ ions. Pure water is (1/10,000,000 N) with respect to H^+ ions at 22°C. In other words the hydrogen ion concentration of pure water is 10^{-7}. It we add some water, the hydrogen ion concentration increases. For example, the concentration may increase to 10. In such a situation OH^- concentration gets reduced to 10^{-12}.

In expressing H^+ ion concentration, the number 10' is always used. Sorensen suggested that the term pH can replace 10^-(p means negative exponent, H means hydrogen). Mathematically pH is defined as the negative logarithm of hydrogen ion concentration. Pure water has 10^{-7} hydrogen ions means its pH is 7. The pH scale ranges from 0 - 14 and can be used to decide the strength of solutions from normal acid to normal base.

Neutral

Acid Base

0 ———————— 7 ———————— 14

Most of the physiological reactions occur at a specific pH. Any alteration of pH will disturb the rate of reaction.

Buffer solutions : Solutions that have a weak acid and its salt or a weak base and its salt are generally stable in that, they resist small changes in H^+ ion concentration due to the addition of acid or base. Such solutions are called buffer solutions and the capacity to maintain their H^+ ion concentration is called *buffer action.*

Acetic acid and sodium acetate are the common buffer pair used in explaining buffer action. If we add NaOH solution to sodium chloride and sodium acetate solutions, the pH of both is increased fast. This means that, the sodium acetate is buffered against an acid but not against a base. If we add NaOH to acetic acid solution, the pH does not change much indicating it is buffered against a base. Hence a mixture of 1M sodium acetate and 1M acetic acid will have buffer action against both acids and bases upto a certain limit.

The two compounds sodium acetate and acetic acid dissociate as shown below

CH_3COONa $CH_3COO^- + Na^+$

CH_3COOH $Ch_3COO^- + H^+$

If we add HCl to the above solution, it (HCI) dissociates into H^+ and Cl^+ ions. H^+ combines with CH_3COO^- to form acetic acid, which dissociates poorly. Cl^- ions combine with Na^+ to form NaCl with the result, there is only a slight change in the hydrogen ion concentration. If we add NaoH, which dissociates into Na^+ and OH^- ions, Na^+ combines with CH_3COO^- to form sodium acetate and OH combines with H^+ present in the solution to form water with the result there is only a slight decrease in H+ ion concentration in the system. All such systems which remove hydrogen or hydroxyl ions from a solution work as buffers.

Buffer solutions are universally present in all living plant cells and play a pivotal role in maintaining the balance of physiological reactions. The biocatalytic activities of enzymes will function only in a limited range of pH and any deviation will completely inhibit their activity.

COLLOIDAL SYSTEMS

Solution in which the particles intermediate in size between those of suspensions and true solutions are distributed throughout is called a colloid. Unlike in a suspension, the particles do not settle after a certain period of standing. For instance a little amount of clay if put into a glass of water and agitated will form a murky soltuion. Soon it cleans out, and particles settle down. But not all particles do so. Some particles still remain dispersed throughout forming a colloidal solution. The suspended particles form the *dispersed phase* and the medium in which the dispersion of particles takes place is called dispersion medium. The dispersed particles are also called *micelles* or *discontinuous phase* and the medium is also called the *intermicellar liquid* or the *continuous medium.*

Phases of colloidal systems : A colloidal system can exist in two states or phases. These may be permanent or the two states may alternate depending on the circumstances. These are the *sol* state and the *gel* state. A colloidal system with more of dispersion medium (high liquidity) is called the *sol* and the one with more of dispersed particles (high amount of solid) is called the *gel* state. It was Thomas Graham (1861) who used these terms, with reference to the liquid and the semiliquid state of the colloids. A *sol can* change into a *gel* and *vice versa*. This is known as *solation* (increase of sol) and *gelation* (increase of gel). Protoplasm is a typical colloid exhibiting both the sol and gel state depending on the circumstances. Such a colloid is called a *reversible colloid.*

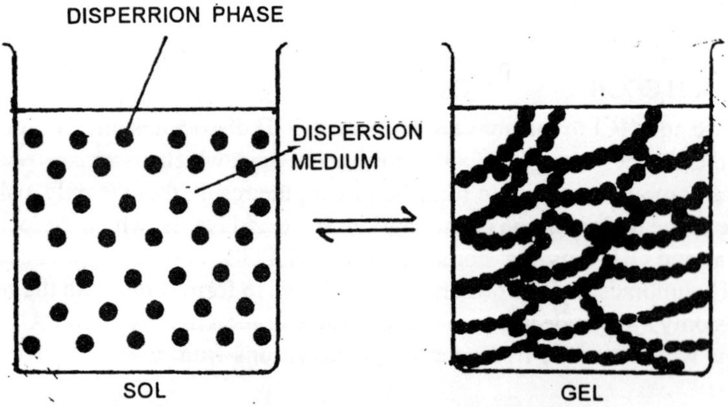

Fig. 3.2 Solutions, Suspensions and Colloids
Sol and Gel forms of a colloidal solution

Types of Colloid systems : There are two general types viz, *Lyophobic* and *Lyophilic*. In the lyophobic system, the dispersed phase and the dispersion medium repel each other, whereas in the lyophilic system they are attracted to each other. When the medium is water, the term hydrophobic and hydrophilic are used to explain the repulsion and attraction between the two components of the colloid. Examples of hydrophobic colloids are metal sulphides and hydroxides. Starch, Agar agar, gum etc., constitute hydrophilic colloids.

Propertieis of colloids : Some of the important properties are-

(a) Filterability : The particles of colloids generally pass through an ordinary filter paper. They cannot however pass through special filters such as parchment paper, sheep bladder membrane etc, whose pore size is smaller. Hence in a mixutre of crystalloids and colloids, the former are filtered out through the membrane while the latter remain where they are. This property is made use of in separating a solution from a colloid and is known as *dialysis*.

(b) Surface phenomenon : The total area of a solid block increases if it is broken into numerous bits as in a colloid. This large surface has added advantage in biological activity. For example, the amount of surface exposed by a solid 1 cm. block is 6 cm . If this cube is divided into 0.1 μ cubes, the amount of surface area exposed would be 6 x 10^5 or an increase of 100,000 times, over the original area. This increases their *adsorption* capacity. Adsorption is defined as the tendency of molecules or ions to adhere to surface of solids or liquids. The amount of adsorption depends on the volume of surface. The molecules absorb to the interface regions of the colloid (the contact area between the two states - *sol* and *gel* is called the interface).

(c) Tyndall effect : When a beam of light is passed through a dark room, it is visible because of the dust particles in air. This is commonly experienced when a beam of sun light enters a dark room through the tiles. The dust particles may be seen moving to and fro. If a beam of light is passed through air which is completely free of any particles, the light is no longer visible. Similarly if a beam of light is passed through a true solution, the course of light cannot be traced. However, a beam of light passed through a colloid will clearly reveal its course because of the suspended particles. The path of light in a colloid can be seen, because of the ability of the particles to absorb light energy and scatter it in the form of short wave length light. This phenonmenon is named *Tyndall effect* and may be used to distinguish a true solution from a colloid.

(d) Brownian movement : This is the to and fro movement of particles in the size range of 4-5μ seen in a colloidal solution. This can be clearly seen when a beam of light passes through a colloidal system. Discovered by Robert Brown in 1828, while observing pollen suspensions in water, its principle was made use of in dark field microscopy later. According to Perrin, the Brownian movement is due to the unequal bombardment of the micells by the intermicellar liquid resulting in oscillations in all directions.

(e) Electrical properties : The colloidal system as such is electrically neutral. But the particles of the dispersed phase are electrically charged, but this charge is neutralized by the opposite charge carried by the intermicellar medium. The *micelles* or particles of any colloid carry always only one type of a charge either negative or positive. When an electrical charge is made to pass through a colloid, the micelles and the intermicellar molecules of the medium migrate to opposite poles. This phenomenon is known as **cataphoresis** or **electrophoresis.**

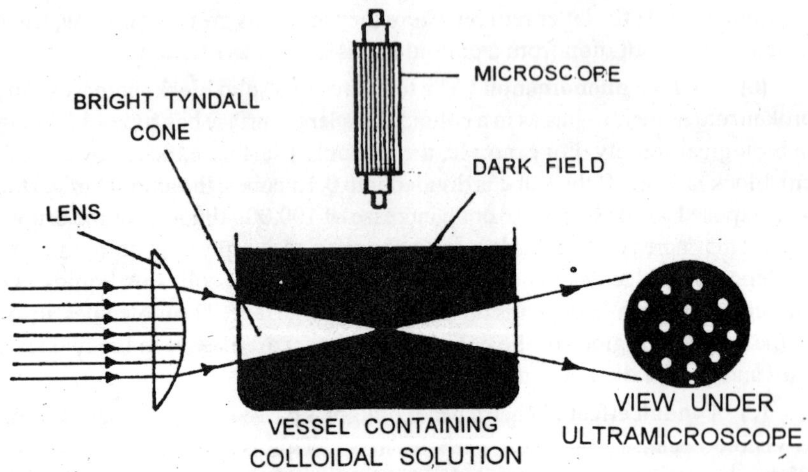

Fig. 3.3 Solutions, Suspensions and Colloids
Tyndal effect in a colloidal solution

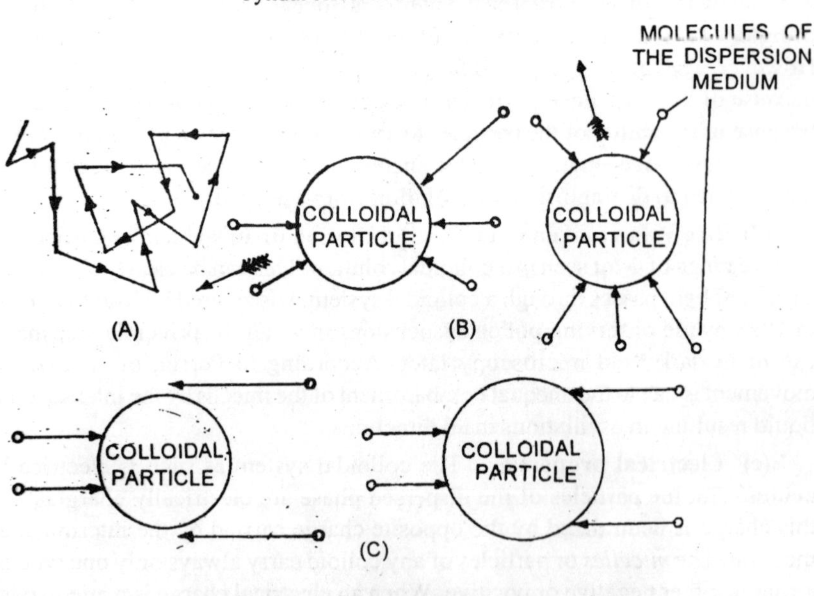

Fig. 3.4 Solutions, Suspensions and Colloids
Brownian movement

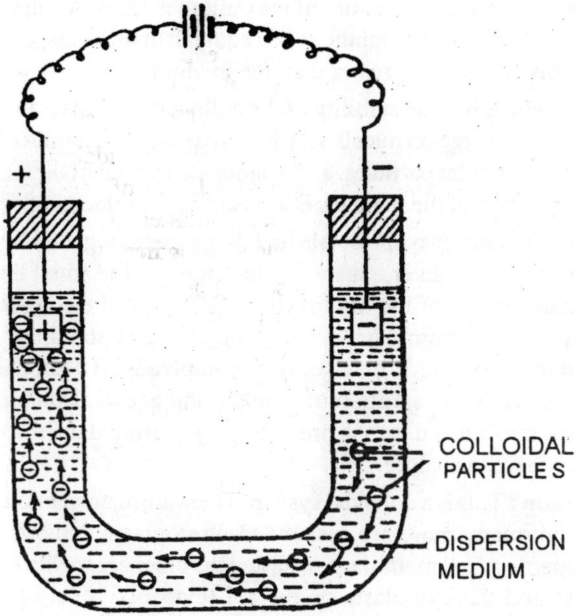

Fig. 3.5 Solutions, Suspensions and Colloids
Cataphoresis in a colloidal system

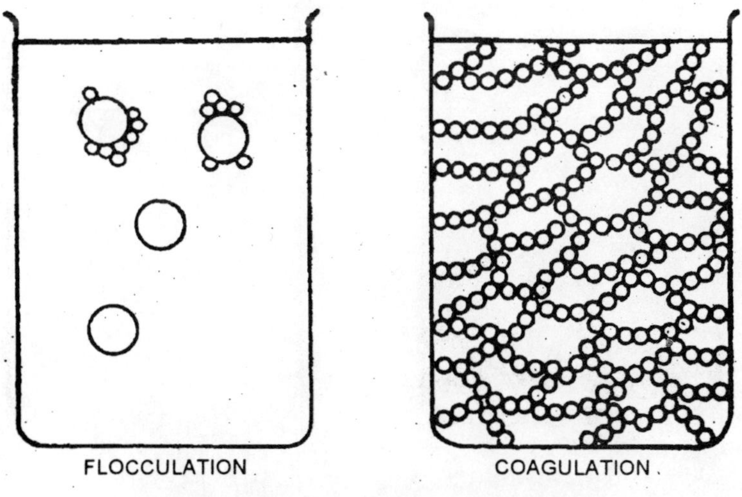

Fig. 3.6 Solutions, Suspensions and Colloids
Flocculation and coagulation in a colloidal system

(f) Viscocity : Viscous nature of the colloid is exhibited in its inability to flow easily. Viscosity of hydrophilic sols is equivalent to the dispersion medium, that of hydrophobic sols is greater than the medium.

(g) Flocculation : The stability of the colloidal systems is maintained due to the presence of, charge on micelles. Addition of small quantities of electrolytes precipitates the colloidal particles, as the latter are robbed of their charge by the former. Precipitation of the micelles is an indication of flocculation.

Emulsions : When two inmiscible liquids are mixed together by vigourously shaking them together, an emulsion results. One can see small droplets of the dispersed phase of one of the liquids, found distributed throughout the other liquid (dispersion medium). After a certain period of standing, the droplets coalesce and the two liquids may separate completely. An emulsion however may be stabilized by the addition of emulsifying agents. Emulsifying agents reduce surface tension and form a protective layer around the micelles making it difficult for them to coalsce.

Protoplasm : This is a colloidal system. There are many particles completely diossolved, but many more are suspended. Protoplasm exhibits many of the colloidal properties. Cell membrane, wall and the other organelles can be thought of as the gel and the cytoplasm as the sol (cytosol). Most of the colloidal properties of the protoplasm are due to the presence of proteins. They are dispersed throughout and offer a wide surface area for enzymatic reactions which are fundamental to life.

4

PERMEABILITY AND MOVEMENT OF SUBSTANCES

A large number of substances enter the cell and an equally large amount goes out of the cell during various biological activities. This is due to the property of *permeability* exhibited by the cell membrane. Membranes are of three categories based on their permeability.

(a) *Permeable* : Allows all substances to enter and exit.

(b) *Semipermeable* : (Selectively permeable, Differentially permeable).

Allows only certain substances, but prevents the entry or exit of certain others.

(c) *Impermeable:* Does not allow either entry or exit of any substance. Acts as a barrier.

In plants, all three types of membranes (barriers) are found. The cell wall is permeable, while the cell membrane is selectively permeable. The cuticle is impermeable. Selective permeability depends on the type of substance, its chemical and physical nature and the nature of the membrane. The movement of solutes is governed by the molecular size, partition coefficient, concentration gradient, electrical charge etc. Permeability of the cell membrane may be altered by changes in the environment of the cell, like ionic concentration, temperature, radiation etc.

Theories of cell perbeability : Several theories have been proposed to account for the differential permeability of the membrane. Some of these are -

(a) Retention pressure theory : This theory was proposed by Traube (1867). According to him. the membrane has affinity towards some substances. These are allowed to diffuse, while others are retained.

(b) Ultrafilteration theory: Proposed by Ruhland, Ruhland and Haffmann (1925), this theory beleives, that the membrane has numerous minute pores acting like *microseive or ultraseive.* This allows substances smaller than the pore size to pass through but bigger ones are held back.

(c) Electrocapillary theory : According to this theory which is a slight modification of the ultrafilteration theory, the *pores* in the membranes have electrical charges negative or positive. Thus permeability depends on the charge of the diffusing particles. Non electrolytes pass through the pores irrespective of the charge in the membrane pore, provided they are smaller in size than that of the pore. Similarly the neutral membrane allows the entry and exit of the electrolytes provided they are smaller in size in comparision with the pore. Positively charged membranes, allow only *anions* while negatively charged membranes, allow only *cations*. In other words, particles carrying the same charge as that of the membrane are repulsed, while those carrying opposite charges are allowed to enter or exit.

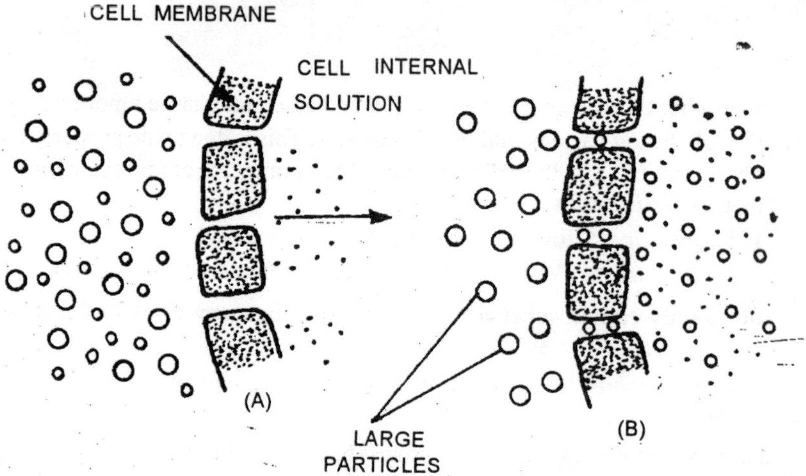

Fig. 4.1 Permeability and Movement
Ultrafilteration theory, **A.** Particles outside the pores, **B.** Smaller particles inside the membrane while larger particles are filtered out.

(d) Fat or Lipoid soluble theory : Since the membranes are made up of lipids, Overtan (1900) opined that, affinity of substances towards lipid greatly helps their mobility across the membrane. The surface of the membrane being made up of lipids, will attract any lipophilic substance and allows it to move. The rate of diffusion .of lipophilic substances is directly proportional to their lipid solubility.

(e) Colloidal theory: Cell membrane and protoplasm are regarded as representing a polyphasic colloid system and the penetration of various substances depends upon the changes in the colloid system. Clowe (1916) regarded it as *phase inversion phenomenon.*

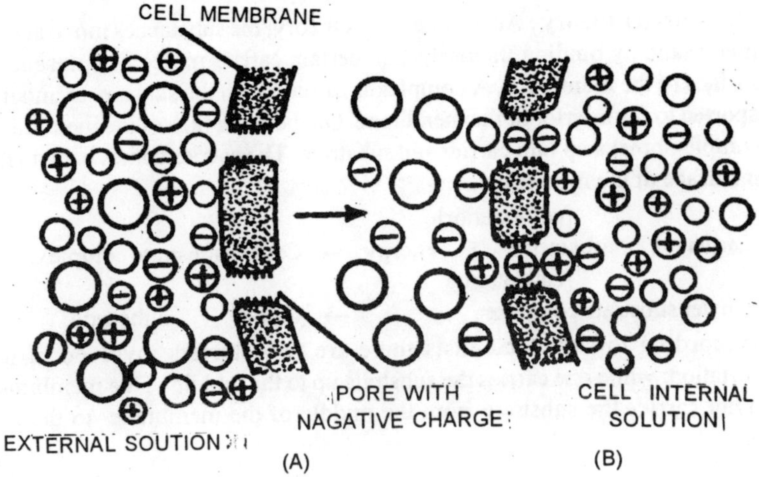

Fig. 4.2 Permeability and Movement
Electrocapillary theory

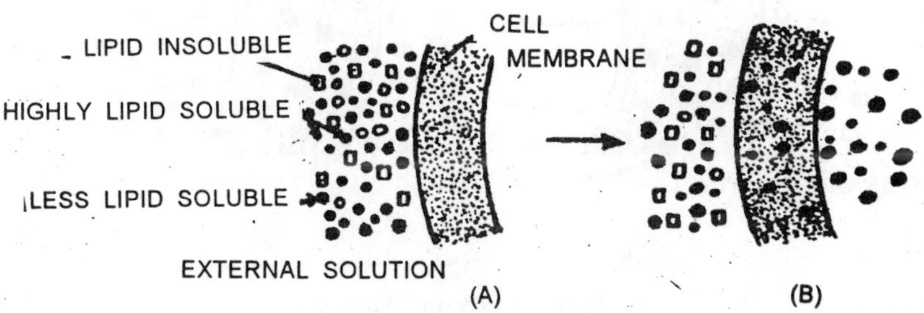

Fig. 4.3 Permeability and Movement
Lipid soluble theory

Objections to the above theories

(a) Particles bigger in size than the pore size, have been found in the cell interior.

(b) Certain solutes are preferentially allowed entry even though they have the identical charge as in the membrane.

(c) Some salts antagonise the entry of others.

(d) Lipid solubility has no effect on the entry or exit of the particles.

(f) Carrier theory : According to this theory, the substances move across the membrane by binding themselves to certain carrier molecules present on the surface of the membrane. A complex is formed with the substances and it is transported to the interior of the membrane. On the inner face of the membrane, the complex breaks up into carrier and substrate. The carrier comes back to the outer surface of the membrane.

[cell exterior]
1. Carrier + substance + energy → Carrier substrate complex.
[cell interior
2. Carrier substrate complex → Carrier + substrate.

According to some scientists there are two carriers involved in the trasportation; while one carries the substrate up to the middle of the membrane, the other carries the substrate from the middle of the membrane to the cell interior.

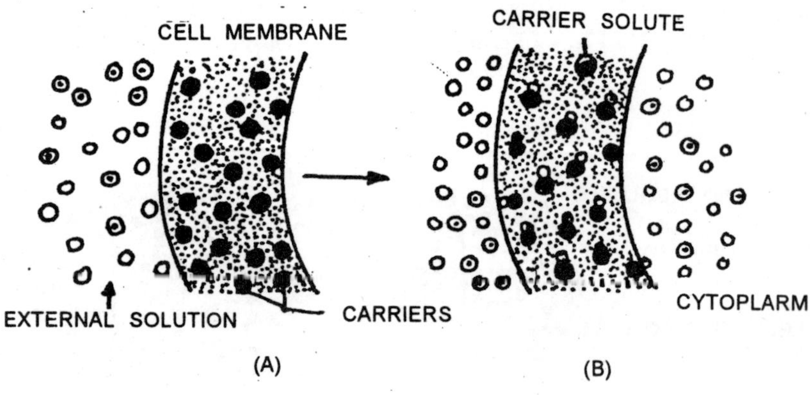

Fig. 4.4 Permeability and Movement
Carrier hypothesis

(g) Carrier vesicle theory : According to Cocking (1965), Stiles and Cockings (1969), the substances are carried across the membrane, by (membrane) vesicles present on the outer surface of the membrane. These carrier vesicles engulf the substances on the membrane exterior and transport it to membrane interior, where the vesicle breaks up releasing the substance. The process is called *pinocytosis.*

MOVEMENT OF SUBSTANCES

In a living cell, which is metabolically active, various substances enter into the cell, and many of them go out of the cell in order to carry on proper biological activities. What is the force that drives these particles or molecules across the membrane? Chiefly there are two categories of forces responsible for the movement of substances across the membrane. These are - the kinetic energy of molecules and the biological energy.

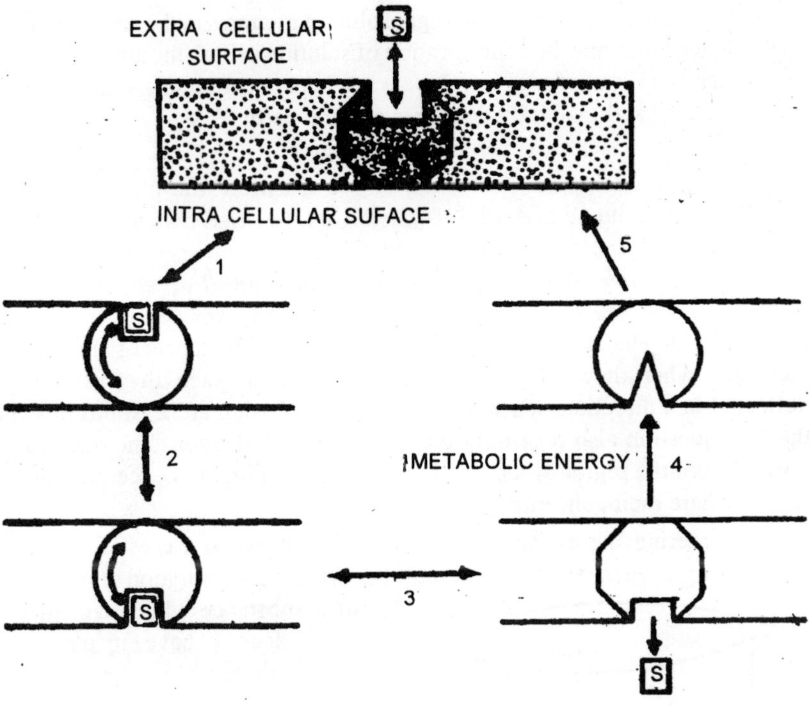

Fig. 4.5 Permeability and Movement
Carrier vesicle theory hypothesis

The kinetic energy is inherent in the molecules. Hence molecules move across the membrane driven by their own inherent energy. This may also be called physical force. The biological energy derived from respiration also drives the molecules across the membrane. Here the molecules themselves have no role to play in their migration. In this chapter we will study the migration of.molecules driven by their own kinetic energy. Chiefly there are three types of movements in this category. These are - Diffusion, Imbibition and Osmosis.

DIFFUSION

Open a bottle of ammonia or light an agarbathi in a room. The fumes or perfume will spread all over the room, and within a short time, even those who are in the fartherest corner of the room can smell it.

Fix a porous partition at the base of a U tube. Fill one arm with water and the other with sugar solution. Slowly sugar solution migrates to the arm containing water. After sometime the concentration of solution in both the arms will be the same. Place a crystal of copper sulphate in a beaker containing water, one can see that the crystal gradually dissolves and the particles start migrating in all directions. The water turns copper sulphate green.

The three examples mentioned above illustrate the phenomenon of diffusion of gas, liquid and solid. *Diffusion maybe defined as the tendency of the molecules of liquid, solid or gas to distribute themselves evenly throughout the available space.* What is the direction of migration? or why do particles migrate? These are the questions to be answered. Answering the second question first- The molecules of any substance have kinetic energy inherent in them and when they collide with one another in a space, they usually get deflected in a direction, where the chances of collision are less; this answers the first question also regarding the direction of deflection. The direction is always from the region where the molecules are in higher concentration to a region where their concentration is less.

Considering the examples mentioned above, when a crystal of copper sulphate is placed in water, it represents the highest concentration of molecules of a substance which can diffuse. Water does not have any molecules of the diffusible substance and hence, the molecules migrate from the crystal in all directions.

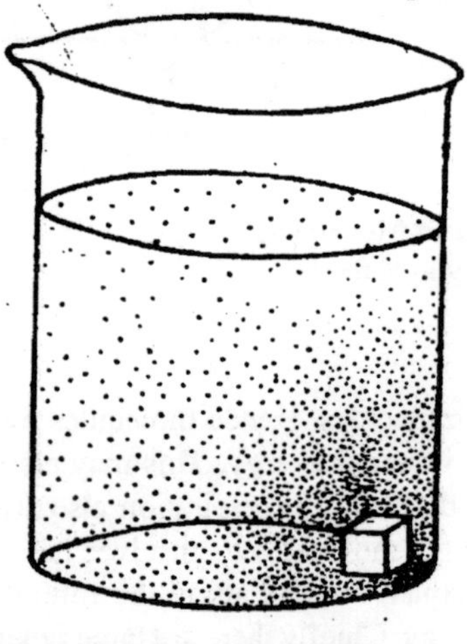

Fig. 4.6 Permeability and Movement
Diffusion of solid particles

Similarly, when a bottle of ammonia is opened in a room, that particular place represents the highest concentration of the diffusible substance Hence molecules of gas diffuse from the region of higher concentration to the region of lesser concentration.

The above examples clearly illustrate like any substance moving in a gradient to the lower point due to gravity, in a concentration gradient also particles move because of their unequal concentration. Is diffusion continuous or will it stop? Taking the above examples once again, when the crystal of copper sulphate is. placed in water, initialy there is one way movement of particles. From the crystal to all the corners of the beaker. However, after the crystal dissolves, there is no migration in any one direction. In fact the rate of migration slows down as the crystal is dissolving. The migration completely stops after the crystal dissolves because equilibrium has been reached. Equilibrium means uniform distribution of molecules throughout the available space; There will be no concentration gradient to force the molecules from one particular direction to the other. This is not to say that there will be no movement at all. The particles move, but not uniformly, hence the net movement is zero, in any direction.

The above explanation gives one more definition for diffusion. It is the *movement of particles from the region of their higher concentration to the region of their lower concentration.*

Diffusion pressure : The particles migrate from the region of their higher concentration to the region of their lower concentration due to a pressure difference in the available space. For instance, when a bottle of ammonia is opened in a room, the bottle represents the highest concentration of ammonia or highest pressure of ammonia molecules. When there is plenty of available place, where there are no molecules of ammonia (least pressure of ammonia molecules), naturally they show a tendency of migration to the region of least pressure. This is diffusion pressure. When there are more number of molecules in a particular place, the particles frequently collide with one another creating a pressure. As a result, if space is available, the molecules migrate. Diffusion pressure may be defined as the pressure exerted by the molecules of liquid, solid or gas (due to their high concentration) to migrate to other areas. Migration naturally is in the direction of least pressure. Hence diffusion may also be defined as *migration of molecules from the region of higher diffusion pressure to the region of lower diffusion pressure.* Diffusion pressure is directly proportional to the concentration of diffusing particles.

Independent diffusion : In any given system, there can be more than one diffusing substance. It may so happen that their concentrations vary and the concentration gradient may be even in opposite directions. How does diffusion take place in such a situation? Does the diffusion of one substance influence the other? The answer is no. The direction of diffusion of a substance is entirely controlled by its concentration gradient and is independent of diffusion pressure of the surrounding substances. Sometimes diffusion may take place simultaneously in same place in opposite directions. This indeed happens in stomata where gaseous exchange takes place during respiration and photosynthesis which take place simultaneously.

The phenomenon mentioned above is called independent diffusion and explains how diffusion of several substances takes place in the same system and simultaneously, independent of one another. In independent diffusion, the diffusion pressure of each substance is called *partial pressure.* In terms of independent diffusion, diffusion may be defined as migration of molecules from the region of their higher partial pressure to the region of their lower partial pressure.

FACTORS AFFECTING DIFFUSION

There are several factors which influence the rate of diffusion. These are -

(a) Temperature : Initially, an increased temperature, increases the rate of diffusion. This is due to the fact that temperature causes increased kinetic energy in particles. This is subject to certain limit. Indefinite increase of temperature will kill the plant.

(b) Density of diffusing substances : Density of the diffusing particles, also affects the rate of diffusion. The rate of diffusion is inversely proportional to the square root of density.

$$\textit{Thus } r = \frac{1}{d}$$

Where r = rate of diffusion of gas, and d = density of gas. This is Graham's law of diffusion. Gases having higher densities will diffuse slowly than gases with lower densities as in diffusion of oxygen and hydrogen.

$$\frac{\gamma H}{\gamma O} = \frac{dO}{dH} = \frac{16}{1} = \frac{4}{1}$$

In the above, hydrogen would diffuse four times faster than oxygen as the density of oxygen is 16 times that of hydrogen.

(c) Concentration of the medium : The rate of diffusion will decrease with the concentration of the medium, and increase with the lowering of density of the medium. A gas diffuses more rapidly in vacuum, than through air.

(d) Diffusion pressure gradient : If the pressure difference is very steep in short distances, diffusion occurs rapidly. If the pressure gradient is gradual, rate of diffusion is slow.

Significance of diffusion in plants

1. Exchange of gases in photosynthesis and repsiration through the same stoma and simultaneously is through independent diffusion.

2. Transpiration from stomata is a process of diffusion of water vapours.
3. Passive uptake of ions is due to diffusion.
4. In the translocation of food material over short distances diffusion plays an important role.

IMBIBITION

When solid substances are kept in liquids, they get soaked and swell in size. It is common knowledge that when seeds are soaked in water, they increase their size due to intake of water. When rubber is soaked in ether, it (rubber) swells due to intake of water. This process of intake of water and subsequent increase in size, noticed in solids is due to a process called imbibition. Imbibition is due to the presence of hydrophilic colloids present on the surface of the imbibant solids which absorb water.

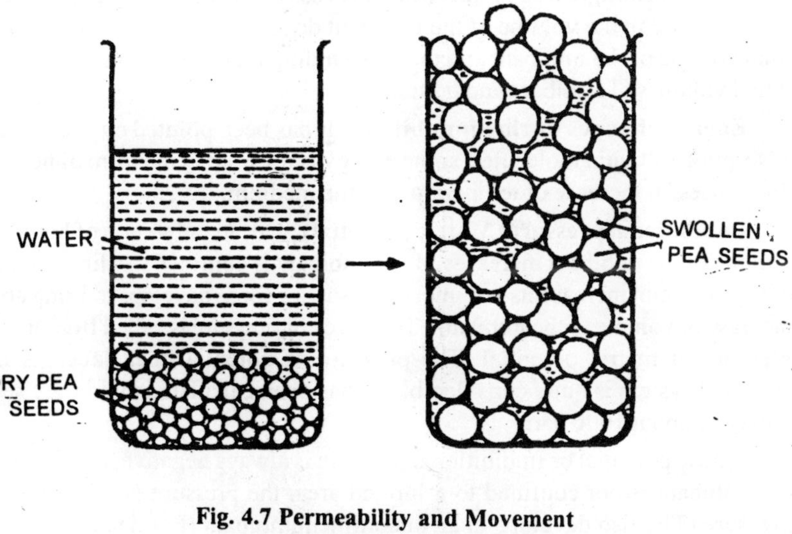

Fig. 4.7 Permeability and Movement
Demonstration of Imbition

There are several day to day examples of imbibition; soaking of gram or pea in water, warping of wooden doors due to imbition of water during rainy season are some of them.

During Imbibition, there is both absorption as well as adsorption. The water gets into the cell, thereby increasing the volume and this is called absorption. Some water molecules however are held over the surface and this is called adsorption. The molecules that are tightly held on the surface of the imbibant lose their kinetic energy due to immobility and this is released from the system in the form of heat. That is why when seeds are soaked, the water will be warm after some time. This is called *heat of wetting*.

Conditions necessary for imbibition : The basic cause of imbibition is a difference in the diffusion pressure of the liquid in the imbibant and the substance imbibed (water). In a dry piece of wood, the water pressure is almost nil and when it is placed in water it readily absorbs water due to a pressure gradient. The imbibition of water continues until the equilibrium i.e. when the pressure of water inside the imbibant equals the pressure of water outside. Another condition necessary for imbibition is affinity between the colloids present on the imbibant's surface and the substance to be imbibed. Thus for example a piece of wood placed in oil does not imbibe, so also a piece of rubber placed in water.

Volume changes due to imbibtion : As has been pointed out already, there is an increase in the volume of the imbibant due to liquid intake. However the final volume of the imbibant and the surrounding liquid is always less than the initial volumes of imbibant and liquid.

Energy changes during imbibition : It has been pointed out, that due to adsorption of liquid molecules, some energy is released in the form of heat and this causes an increase of temperature in the medium.

Imbibitional pressure (Matric potential) : Due to imbibition of liquid, the volume of the imbibant increases, if the imbibant is confined to a limited space where it cannot increase its volume, a pressure or force develops to bring about increase of volume in the imbibant. This force or pressure is called Imbibitional pressure or matric potential. The pressure of potential that develops in a confined system is quite considerable. It has been calculated that in dry seeds it may be about 1000 bars.

Matric potential or Imbibitional pressure is always negative, in the sense if the imbibant is not confined to a limited area, the pressure is nil. No turgor pressure (TP) also develops in an unconfined imbibant (For details of TP see Osmosis). The relationship between the matric potential and the pressure or potential of water is equal and it is expressed in the form of following mathematical expression.

$$\underset{\text{(Water potential)}}{\psi w} = \underset{\text{(Matric potential)}}{\psi m}$$

Measurement of Imbibitional pressure (IP) : This can be done by two methods; Direct method and osmotic method.

In direct method, the pressure is measured in terms of the counter pressure that is needed to be applied in order not to allow IP to develop. In other words an opposing pressure is applied (which can be meausred) to prevent increase in volume of the imbibant. With the help of the following experiment IP can be measured.

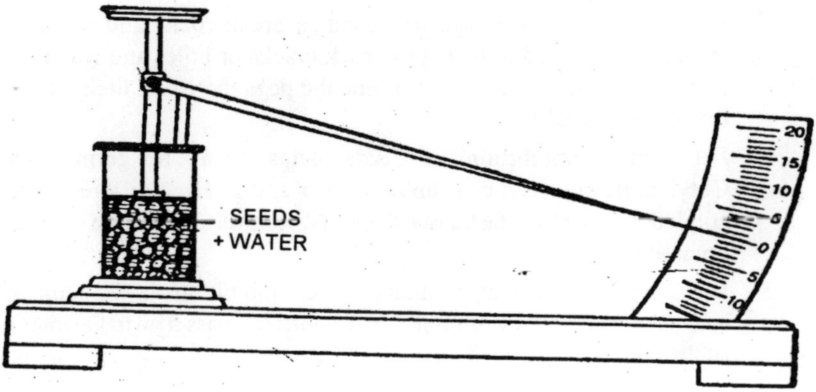

Fig. 4.8 Permeability and Movement
Demonstration of Imbitional pressure

A few gram seeds are soaked in water and kept in ajar. A disc is placed above the level of the seeds in the jar. With the help of a split cork the other end of the disc is attached through a rod to a pointer, which moves on a semilunar scale. Initially the pointer is adjusted to zero, the apparatus is made airtight and left standing for about 24 hours. Afterwards it is discovered that the disk has been pushed up and the pointer has moved away from zero. This gives an idea of the imbibitional pressure developed by the seeds.

Factors affecting imbibition

1. Temperature : Initially, the rate of imbibition increases with increasing temperature of the imbibant. Increase of temperature has an effect of increase of kinetic energy of imbibant molecules.

2. Concentration of solute in the medium : The rate of imbibition decreases, with the increased concentration of solutes in the medium. This is due to the decreased water pressure in a solvent consisting of solutes. Addition of solvent to a medium increases imbition by increasing water pressure.

3. Effect of pH : The pH of the medium also influences the rate of imbibition.

4. Specificity : The imbibants are specific to the liquids they imbibe. In the absence of the specific liquid, imbibition will not take place.

Significance of Imbibition

1. In olden days, imbibition was used to break rocks and boulders. Wooden pegs used to be kept in rock cracks or holes and water was poured over them. In course of time the pegs increased their volume breaking the rocks.
2. Differential imbibition of seeds helps them in germination. Cotyledons and seed coat imbibe water and swell at different rates; cotyledons swell at a faster rate than seed coat, resulting in the breaking open of the latter.
3. In the initial stages of absorption of water, imbibition plays a dominant role. Root hair cell walls imbibe water, and then pass it on to the interior of the cell.

OSMOSIS

This is another kind of physical force of the molecules governing the entry and exit of the substances with reference to a cell. Osmosis may be regarded as a special type of diffusion, involving the migration of only the solvent molecules across a differentially permeable membrane.

Discovery : Nollet (1748) discovered the phenomenon of osmosis using a pig's bladder. He tied the membrane over a jar, filled with wine and immersed it in a glass trough containing water. After some time. he noticed that, the level of the wine rose in the jar due to the entry of water. Dutrochet, later demonstrated it using animal membranes.

Physical demonstration of Osmosis : Take a beaker half filled with water. Take a thistle funnel and tie its mouth with a parchment paper, sheep bladder or egg membrane etc. These membranes are selectively permeable since they allow only certain particles to pass through and not all. In this case, the membrane is permeable to water only. Fix the thistle funnel to a stand and immerse it in the beaker containing water. Fill the thistle funnel with 5% sugar solution and note the level. After sometime it may be observed that the level of the sugar solution in the thistle funnel rises. Why? Water from the beaker must have entered into the thistle funnel. Now taste the water in the beaker. It does not taste sweet, because the membrane has allowed only the water molecules to enter into the thistle funnel, but has not allowed the sugar molecules into the water contained in the beaker. Water migrates into the thistle funnel, because it is in higher concentration in the beaker. Hence *osmosis may be defined as the migration of* *solvent molecules from the region of thier higher concentration, to the region of their lower concentration through a selectively permeable membrane.*

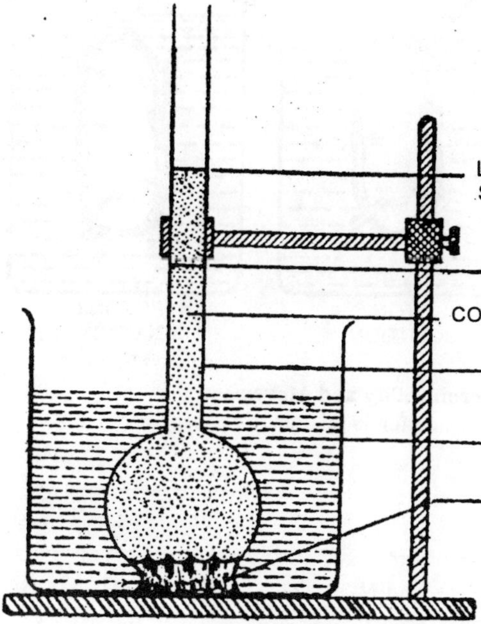

Fig. 4.9 Permeability and Movement
Physical demonstration of osmosis

This is not to say that osmosis is observed o. ly when a solution and solvent are involved. Osmosis may take place between two solutions also depending on the concentration gradient. If separate two solutions (solvent systems) one of 5% (A) and ano..ler of 10% (B) by means of a differentially permeable membrane, we notice that solvent migrates from A to B. The reason being 5% solution is a stronger solvent when compared with 10% solution. In 10% solution the amount of solvent is less than what is present in 5% solution. (If a 10% and 5% solution are separated by means of a permeable membrane, the solution migrates from B to A. Whereas in osmosis, only the solvent migrates from A to B along its concentration gradient)

Concentration of solvents : Pure water is the strongest solvent.. The addition of solute to a solvent turns it into a solution. With the increased concentration of solution, the solvent concentration goes down. *The concentration of the solvent and solution are inversely proportional.*

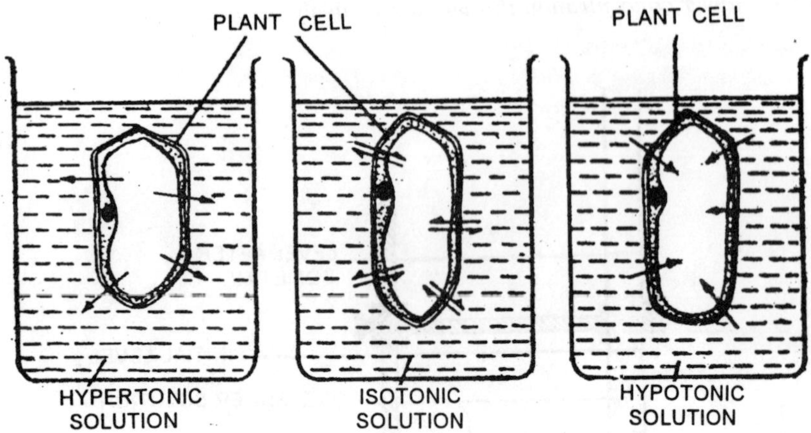

Fig. 4.10 Permeability and Movement
Concentration of solution in relation to cell sap

Solution Concentration (%)	0	5	10	15	20
Solvent Concentration (%)	100	95	90	85	80

Three types of solutions are recognised with reference to concentration and osmotic migration.

Hypertonic : This is a strong solution (weak solvent) and as such it attracts solvent from other solutions.

Hypotonic : This is a weak solution (strong solvent), as such it loses water to other solutions.

Isotonic : The concentration of the solution is such that it neither gains nor loses water in an osmotic system.

It has to be noted that the terms hyper, hypo and isotomic solutions do not have an absolute meaning. It is only relative and can be used only in comparison with another solution. For example a 5% solution is hypertonic with reference to a 2% solution, but hypotonic with reference to a 10% solution. In practise however, the three terms are used with reference to cell sap concentration. If a cell loses water when kept in a solution it (solution) is said to be hypertonic and if it (cell) attracts water from the solution, the solution is said to be hypotonic. If a cell neither gains nor loses water, the solution is said to be isotonic. This is made use of in deciding the cell sap concentration.

Physiological demonstration of Osmosis : Osmatic phenomenon can be demonstrated both by abiological and biological systems, whenever two solvent systems are separated by means of a differentially permeable membrane. This is of great significance in plants as it governs the entry and exit of water into and from the plants. Plant cell in fact acts as an osmotic system. This may be demonstrated with the help of a potato osmoscope.

Take a potato and cut it into two pieces. Peel of the skin and scoop some amount of tissue from the centre and shape it into a cup. Prepare two cups of potato. With the help of this osmosis can be demonstrated, hence it is called an osmoscope. Fill one osmoscope (A) with 3% sugar solution and note the level. (A small pin can be fixed to note the level). The other osmoscope (B) is filled with water and the level is noted. Place A in a petri dish containing water and B in a petri dish containing 3% sugar solution. Allow the process of osmosis to go on for some time. After some time one can notice that the level of sugar solution has gone up in A, while the level of water has gone down in B. Due to osmotic entry of water, the level of the sugar solution goes up, while in B, water goes out into the petri dish as there is a hypertonic solution. Both these illustrate the phenomenon of osmosis. However if we compare the potato cup to a confined system or cell in A it is *endosmosis* (entry of water into a closed system) and in B it is *exosmosis* (exit of water from a closed system)

Osmotic pressure (osmotic potential)

When two solvent systems of different concentrations are separated by means of a selectively permeable membrane, solvent migrates from the region of its higher concentration to the region of its lower concentration due to osmosis. Both these solvent systems may be termed osmotic systems. A strong solvent migrates to a weak solvent. This mean that the weak solvent has the capacity or attraction or force to draw water from a strong solvent (a weak solution). Since a weak solvent is a strong solution, it has less solvent and so can attract water from a weak solution (strong solvent).

This force or pressure that is there in a strong solution is called its osmotic force or pressure. It is due to this osmotic force that a strong solution draws water from a weak solution. However a more correct definition of osmotic pressure (OP) should be explained in negative terms, because OP cannot be seen or demonstrated directly. When the strong solution draws water from a weak solution it does so with some force or pressure. It we apply a counter pressure on an osmotic system so that it cannot attract water, that counter pressure will be equivalent to OP. Hence OP may be defined as the *pressure that is applied on an active osmotic system in order to prevent the entry of water into it.* Initially when the pressure is applied, entry of water into an osmotic system may continue. But at a particular magnitude of the pressure osmosis would stop and that is the OP of that system. Strong solutions (weak solvents) have higher OP ie. more amount of pressure is needed to prevent the entry of water into that system. OP is directly proportional to the number of solute molecules per solvent molecules. Hence OP may also be defined as "excess hydrostatic pressure which must be applied on a solution in order to make its solvent potential equal to that of a pure solvent".

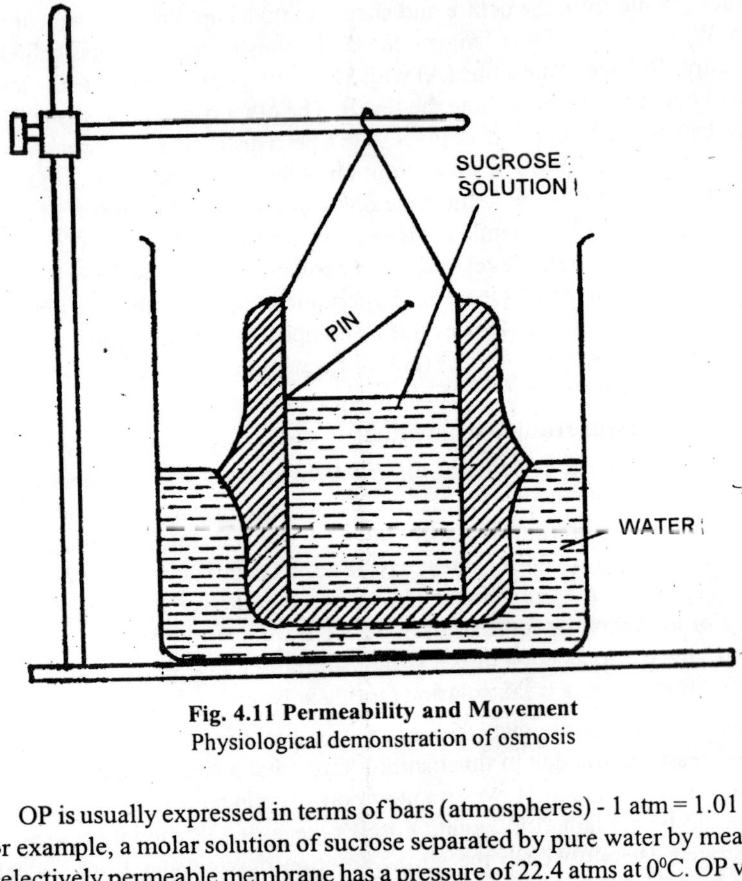

Fig. 4.11 Permeability and Movement
Physiological demonstration of osmosis

OP is usually expressed in terms of bars (atmospheres) - 1 atm = 1.01 bar). For example, a molar solution of sucrose separated by pure water by means of a selectively permeable membrane has a pressure of 22.4 atms at 0°C. OP varies with the temperature of the medium. At lower temperatures it tends to be low.

OP is usually denoted by the symbol *n*. Vant Hoff formulated an equation to relate OP to concentration of the solute in the solution. This is as follows

$$V = n_s RT$$

Here = is OP, V = volume of the solvent, ns = moles of the solute, R = gas constant (0.0820 litre atm/degree mole), T = absolute temperature.

(Note : This equation is not applicable to electrolytes)

OP is evident only in the case of a confined solvent. But in all situations the solvents are not confined. In such a situation, instead of using the term OP, the term osmotic potential is used. Both are equal and synonymus in the sense, that they refer to the same phenomenon. Osmotic potential seems to be a more appropriate term since it is not a pressure but the property of a solution. Osmotic potential is numerically equal to the OP, but is negative in sign. In a pure solvent osmotic potential is zero, because it is saturated with the solvent.

Measurement of osmotic pressure : Many attempts have been made since a long time to measure the osmotic pressure in a system. Maurice Traube (1867) was the first to prepare an artificial semipermeable membrane and showed the physiological significance of OP in plant cells. One of the membranes was prepared by taking a drop of copper sulphate at the end of a glass rod, which was immersed in a solution of potassium ferrocyanide. A membrane or cell was formed at the end of the rod. This membrane was too fragile and could not withstand the pressure. Pfeffer (1877) a German botanist improved upon Traube's work and prepared an osmometre to measure OP.

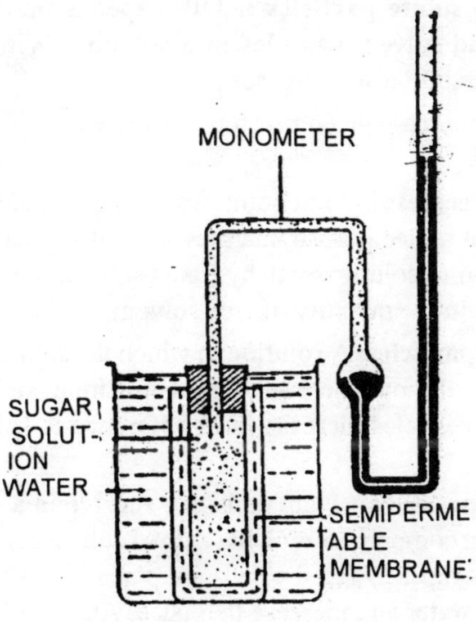

Fig. 4.12 Permeability and Movement
Pfeffers Osmometer

Pfeffer's osmometre : This apparatus, developed by Pfeffer is used for the measurement of osmotic pressure. The apparatus basically consists of a *Pfeffer's cell* and a *mercury manometre.* Pfeffer's cell can be prepared as follows. Take an unglazed porous (Porcelain) pot and wash it. After drying, fill it with a solution of copper sulphate and place it in an outer container containing potassium ferrocyanide solution. The two solutions migrate in opposite directions and meet inside the wall of the porous pot. As a result, a membrane of copper ferrocyanide is formed inside the wall of the pot. This membrane is semipermeable. After some time, remove the pot, pour out the copper sulphate solution and wash it (pot) thoroughly. This is the Pfeffer's cell with a membrane of copper ferrocyanide inside the wall. Fill the Pfeffer's cell with sugar solution and close it by a cork through which passes a glass tube connected to a mercury manometre (as shown in the figure). The Pfeffer's cell is now placed in a glass trough containing water. Water from the trough enters into the Pfeffer's cell raising its level. This in turn raises the level of mercury in the manometre. OP can be measured in terms of the increase in mercury manometre.

In addition to the above, two more methods are employed to measure OP in plant cells. These are explained under plasmolysis.

Factors affecting OP

1. **Concentration of the solute particles :** OP depends upon the proportion of solute particles and solvent particles in a solution. With the increased concentration of solutes, OP also increases.

2. **Temperature:** Initially, increase in temperature, increases OP. This is however subject to a certain limit.

3. **Hydration of solute molecules:** In some solutions, solute particles are intimately associated with solvent molecules, so that they are not available to increase the solvent concentration. A solution with hydrated solute molecules has a higher OP due to reduction in the quantity of free solvent.

4. **lonization of the solute particles:** A solution in which the solutes are ionized has a higher OP than in a solution where solutes are non ionized. With ionization, the number of solute particles increase (with reference to solvent particles) thus increasing the OP.

Turgor pressure and wall pressure : Plant cells can survive in a wide range of osmotic concentrations in comparison with the animal cell, because of the presence of cell wall which is rigid. When the cells are placed in water or hypotonic solutions, they absorb water and increase their size. As a result, the elastic plasma membrane presses the cell wall. The pressure that it built up by the cytoplasmic membrane is called the *turgor pressure.* Turgor pressure will be maximum when the cell is not flaccid ie. when solvent concentration is maximum inside the cell. When the cell has less solvent (as it happens when it is placed in a medium of hypertonic solution), the membrane will not press the cell wall and turgor pressure is lowest at that time. When the cell membrane is applying a pressure, the cell wall being rigid, (in order not to burst and retain the shape), also applies an equal amount of pressure in the opposite direction. This is known as the *wall pressure.* As a result of these two pressures the cells will be turgid. At any given time turgor pressure and wall pressure are equal (TP = WP)

Water potential : Potential may be defined as the capacity to do the work. When a chemical system does some work at a constant temperature, the free energy falls down. Thus the free energy represents the potential of the system to do the work. The free energy per molecule in any chemical system is called its chemical potential. The chemical potential of water is called *water potential.* Water potential in any solvent system actually represents the difference between the chemical potential of water in that system and pure water. Water potential is always represented by a negative sign. Thus the water potential of pure water is zero. The greek letter psi (ψ) designates the water potential in a system. It is expressed in terms bars or megapascals (Mpa = 10 bars). Water potential in a system can be calculated as follows.

$$\psi = (RT \text{ in } e/e^0) / V$$

Here ψ = water potential in the system in bars, R = ideal gas constant (8.3×10^{-1} ergs mol^{-1} deg^{-1}), T = absolute temperature in ^0K, e= vapour pressure of water in the system at T, e^0 = vapour pressure of pure water at T and V == molal volume of water (18.0 cm^3 mol^{-1}).

Water potential is a diagnostic character of the status of water in a cell or tissue. The lower the potential, the greater is the capacity to absorb water. Water potential is made use of in measuring *water deficit* and *water stress.* In plant tissues water potential is always less than 0 bar. Leaves and aerial parts of plants in soils with adequate moisture supply will normally have a water potential between 2 and 8 bars. With decreasing supply of water, the values become more negative and at 15 bars plant tissues cease to grow.

Diffusion pressure deficit (DPD) : Molecules of liquids, solids and gas have a pressure to diffuse when present in a particular concentration. In the case of a pure solvent, there is maximum diffusion pressure of the solvent molecules, but with the addition of solutes, the pressure of the solvent molecules comes down. The amount by which the diffusion pressure of solvent molecules in a solution is lower than that of the pure solvent at identical temperature and atmospheric pressure is called *Diffusion pressure deficit.* DPD is increased by the addition (DP is reduced) of solutes into a solvent system.

The terms DP and DPD were introduced by B.S. Meyer (1938). These were originally termed as suction pressure or suction force by Renner (1915). Nowadays however the term water potential has replaced DPD since both refdine the same phenomenon. While DPD indicates a positive value, water potential indicates a negative value. In order to explain the migration of solvent molecules along a concentration gradient, the term DPD seems to be inappropriate, since it explains solvent migration from lower DPD to higher DPD ie. up a gradient. While the migration of water from lower puressure deficit to higher pressure deficit is true, the exact explanation should be from higher pressure to a lower pressure *ie.* down a gradient. From this point of view water pontential is a better term. There are three major components of water potential in a plant cell. These are-

(a) matric potential,
(b) solute pontential and
(c) pressure potential.

Water potential in a plant is the sum total of all these three.

Matric potential represents the surface of the cell, soil etc., where hydrophilic colloids are present and water molecules get adsorbed to them. The matric potential (Vm) is generally not significant in osmotic movement because it does not allow the free movement of water molecues.

Solute potential (ψ s) is also known as the osmotic potential. It may be defined as the amount by which water potential is reduced as a result of the presence of solute particles. Solute potential is always represented in negative numbers (ie. so much quantity of solvent is absent in comparison with a pure solvent)

Pressure potential (ψ p) is represented by the cell wall and cell membrane. The cell wall exerts a pressure (WP) on the cell contents. Similarly the cell membrane also exerts a pressure (TP) on the cell wall by virtue of its solvent content. The pressure potential is positive in that it denotes the amount of moisture already present. Hence water potential may be represented symbolically as follows

$$\psi = \psi m + \psi s + \psi p.$$

Relationship between OP, TP and DPD

When a solution of a particular concentration present in A is separated from pure water present in B by means of a semipermeable membrane, because of lower DPD of solvent in B, solvent migrates to A. The solution in A has some amount of OP. At this stage OP and DPD are approximately equal. It will be completely equal when there is absolutely no solvent in A. The amount of solvent in A is represented by TP. Hence A can receive only so much quantity of water (because of DPD) from B to reach equillibrium after subtracting whatever solvent that is already there (TP).

Hence

DPD = OP (approximately)

DPD = OP-TP (= WP)

Plant cell as on osmotic system : Plant cells and tissues function like osmoscopes and exhibit osmotic phenomenon. When cells are placed in the medium of a solvent system, they react to it and either draw or give out water depending on the concentration of the external medium. Various pressures (potentials) connected with osmosis like OP, TP (= WP), and DPD interact. The interactions between these cells is called the osmotic relationship of the cell.

A typical plant cell has the cellulosic cell wall as the outer boundary. The cell wall is porous and permeable and elastic to some extent. Interior to this is the plasmalemma or the plasma membrane which is selectively permeable. Generally there will be a vacuole within the plant cell whose membrane is called the tonoplast. Cytoplasm forms a matrix between the plasma membrane and the tonoplast. The vacuole consists of a solvent system viz. the cell sap. Thus the cell has all the structural attributes of an osmotic system.

When a plant cell is kept in pure water, due to difference in concentration between the cell interior and exterior, water moves into the cell and the flaccid cell becomes turgid due to intake of water. Turgor pressure increases as the plasma membrane applies pressure on the cell wall. The cell wall under tension presses the membrane in opposite direction (wall pressure). Continued entry of water increases the TP. Finally entry of water stops when equilibrium is reached. At this point, the force of entry of water is equalled by the force of exit, hence no migration of water (OP = TP) takes place.

Initially, water migrates into the cell due to the concentration gradient. The cytoplasm and cell sap represent concentrated solutions, when compared with the external medium. When the cell is kept in a hypertonic solution however, there is exit of water from the cell rather than entry into it (see plasmolysis).

PLASMOLYSIS *(Plasma* = cytoplasm, *lysis* = breakdown) : When a plant or animal cell is kept in a hypertonic solution, water from cell sap flows out due to a reverse concentration gradient (concentration gradient pointing out of the cell). As the solution in the external medium is more concentrated, it has more OP and hence water moves out of the cell resulting in shrinkage of the cell volume. If the conditions persist for a long time the cell will die. This phenomenon of water loss from cells which is akin to exosmosis is called plasmolysis. Wilting of plants noticed under conditions of water scarcity is an indication of plasmolysis of cells. There are three stages in plasmolysis - viz. Incipient plasmolysis, evident plasmolysis and final plasmolysis.

Incipient plasmolysis : In a turgid cell, with adequate amount of water, the plasma membrane closely appresses the cell wall and is in contact with it alround. When this cell is kept in a hypertonic solution, water starts moving out of the cell. Initially there will be no effect on the cell wall. But as water continues to go out, the cell shrinks in volume. But the plasma membrane maintains its contact with the cell wall alround due to its elastic capacity. As the exit of water continues, the plasma membrane reaches its limit of elasticity and gets torn from the cell wall at the corners, while maintaining contact at other regions. This is the first stage of plasmolysis called incipient plasmolysis.

Evident plasmolysis : The cell under hypertonic conditions continues to lose water to the external medium and shrinks further in volume. The plasma membrane gets torn completely from the cell wall and goes on contracting. This is the second stage of plasmolysis called evident plasmolysis.

Final plasmolysis : As exosmosis continues, the shrinkage of the cell and cytoplasm reaches the minimum limit and no further shrinkage in volume is possible. Cytoplasm will be completely free from the cell wall and remains in the centre of the cell. Based on the final shape of the cytoplasm the final plasmolysis is divided into two types *concave plasmolysis* and *convex plasmolysis*. In the former, cytoplasm assumes a concave shape and in the latter it assumes a convex shape. A cell which has undergone plasmolysis is called *plasmolysed*.

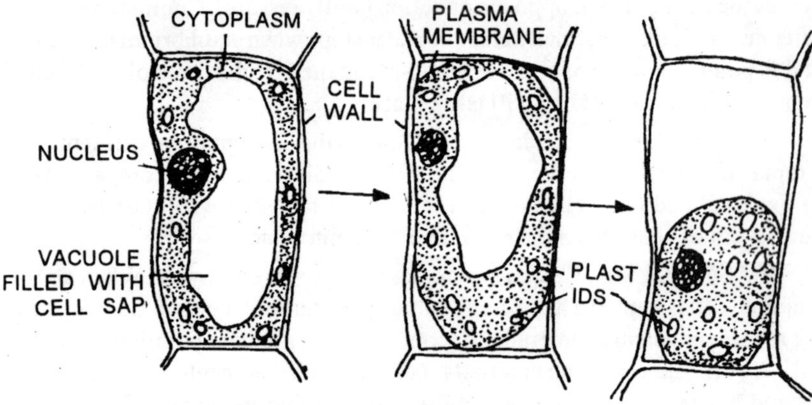

Fig. 4.13 Permeability and Movement
Plasmolysis of a plant cell

Deplasmolysis : A plasmolysed cell can be revived, by transfering it into a medium of water or hypotonic solution. On such a transfer plasmolysis is reversed and the cell becomes turgid once again due to the entry of water. It regains its original shape and size. This phenomenon of revival of a plasmolysed cell is called deplasmolysis.

Measurement of OP by plasmolytic method

The plasmolytic method of determining the OP of plant cells was first devised by Devries (1884). The method involves finding out the OP of the cell sap by keeping the cell in different concentrations of solution (from hypertonic to hypotonic). In one particular concentration of the external solution, there is neither endosmosis nor exosmosis. In this concentration there is equilibrium and the forces of entry and exit of water are equal. The concentration of the cell sap is approximately equal to the concentration of the external medium at this stage. The word approximately is used here deliberately, because when a cell is kept in a solution some shrinkage takes place before plasmolysis actually takes place. Usually OP obtained by this method is slightly higher than the actual OP.

A slight modification of the plasmolytic method consists in measuring the length, breadth and volume of the cells before and after immersion in solutions of known osmotic pressure. OP is then calculated using the formula of Hofler (1917).

$$p = \frac{P_1 \times V_1}{v}$$

Where p = OP of cell, v = volume of cell before immersion, V_1 = volume of cell after immersion and P_1 = OP of external medium.

Significance of osmosis

1. Absorption of water in large quantities is due to osmosis.
2. Distribution and movement of water between the cells of various parts of the plant body is due to osmosis.
3. Various plant organs maintain their stature, posture, shape, size etc., due to turgidity of their cells. Osmosis plays an important role in maintaining the turgidity of the cells.
4. Turgor movement of the guard cells helps in the opening and closing of the stomata.
5. Certain plant movements (leaflets of *touch me not* plants) like turgor movements are due to osmotic changes.
6. The exposure of the leaf to sun light is due to turgidity of its cells, which is again due to osmosis.
7. Growing points should have turgid cells both at shoot tip and at root tip. Turgid cells of the root tip easily penetrate the soil particles.
8. Resistance of plants to drought and frost is brought about by their high osmotic concentration.
9. Mechanical bursting of fruits *(Balsam, Ecballium* etc), and dehiscence of sporangia (in many ferns) is due to turgor pressure.

5

ABSORPTION OF WATER

Water is one of the most important compounds that profoundly influences life. In fact it is impossible to visualize life in its myriads of forms in dehydrated state. It is widely believed that life itself originated in some quiet corner of the primordial ocean (on the young Earth) in the remote past. There is no wonder then water is aptly called the *liquid of life* or the *elixir of life*. From the point of view of plant life, their very sustenance is dependent on water. The following is a list of the roles played by water in the life of plants.

1. Water is an essential and major constituent of protoplasm. In fact bulk of protoplasm is water. All the cell organelles maintain their structure and function in the presence of adequate quantity of water.
2. Distribution of various dissolved substances from one cell to another and from one part to another can take place only in the medium of water.
3. Water is an excellent solvent. It also acts as a reagent in various reactions in plant cells. In photosynthesis and respiration water plays an important role.
4. Breakdown of various chemical compounds takes place only in the medium of water. During germination, complex food substances' stored in the seed breakdown into simple constituents. This is possible only when water is present.
5. During metabolic reactions, the inner temperature of the plant rises. Water stream flowing throughout the plant body acts as a coolant and maintains the temperature.
6. Adequate amount of water is necessary in cells and tissues to maintain the turgidity of the plant body.
7. Enzymes, without whose mediation, virtually, no biological reaction can take place, function only in a hydrated medium.
8. Plants take in their nutrients from soil only in a solution form. Soil water thus is very essential for the survival of the plants.

9. A network of thin layer of water surrounding each cell plays an importnat role in the movement of dissolved substances.
10. In lower plants water is essential for the mobility of gametes.
11. Dehiscence of sporangia and dispersal of spores (hygroscopic reaction) is aided by atmospheric water.
12. Water also helps in pollination and dispersal of fruits in many of the higher plants.

Composition and properties of water

Chemically, each molecule of water is made up of two atoms of hydrogen and one atom of oxygen (H_2O). Hydrogen atom has one electron, while oxygen atom has eight electrons. When chemical combination occurs, there is a sharing of electrons in pairs between oxygen and hydrogen atoms. This type of bond is known as the covalent bond in which each atom contributes one electron and these are held jointly by both the nuclei. Water molecules are highly stable because of covalent bonds. Water molecules have an approximate diameter of 0.25 nm.

Evidence indicates that the H and 0 nuclei in a water molecule have an asymmetrical configuration as a result of which the oxygen side is more negative than the hydrogen side. Hence water molecule is said to be a *dipole.* Water molecules form a special kind of bond called hydrogen bond, which refers to the attraction of a hydrogen nucleus by two other atoms so that acting as a link between them. It is this hydrogen bonding which is responsible for the strong interaction of water molecules among them and with other ions.

Physical and chemical properties

1. Liquid water is colourless, odourless and tasteless.
2. It expands on freezing, while all other substances contract. This is the reason why ice floats on water on accout of its lesser density.
3. The molecular weight of water is low (18), but retains the liquid form because of hydrogen bonds.
4. Water has a high boiling point (100°C) and high heat of vapourization (540 cal g^{-1}' at 100°C). High heat of vapourization has a cooling effect in evaporation, as it takes away energy in the form of heat.
5. The specific heat of water is the highest save for liquid ammonia. It has a great capacity to absorb heat with very little rise in temperature. More heat is required to raise the temperature of 1 gm of water by 1°C than for any other substance. The high specific heat of water helps plants maintain stable internal temperature.
6. Water molecules have a high *cohesion force.* They have the property

to cling to others tightly. It is this tensile strength that maintains a continuous stream of water inside the plant.

7. Water has the highest surface tension except for mercury. Molecules of water at the surface of liquid are attracted more strongly by the molecules in the interior of the liquid, than by the molecules of air above the surface of water. As a result, the liquid air interface behaves like an elastic sheet. This property is Called *surface tension.* It is this surface tension that enables water to move by capillarity between soil micelles and plant cells.
8. Water has an affinity or adhesive property to molecules of many other substances.
9. Water acts as a solvent to many substances than any other liquid. Electrolytes that dissolve in water have their positively and negatively charged ions to attract to the positive and negative sides of water molecules respectively. Non electrolytes also dissolve readily in water, as these molecules form hydrogen bonding with water molecules. The ionization of electrolytes in water makes it a good conductor of electricity.
10. Pure water is generally transparent and allows the transmission of visible light.

Availability of water to the plants : Water is available to the plants both in atmosphere and soil But the quantity of water present in the atmosphere is woefully less for the requirement of plants. A large majority of plants absorb water from soil (except for rootless aquatic plants).

Soil plays a dominant role in the life of terrestrial plants by providing them with nutrition and water. It also provides anchorage for the plants. The chief source of soil water is rainfall; besides this, melting of snow also provides water. According to its availability to plants, soil water is described in three different terms.

Holard - Total water present in the soil.

Chesard - Soil water available to the plants.

Echard - Soil water which cannot be absorbed by plants.

In another classification. Five types of soil water are recognized based on their availability to the plants. These are -

(i) **Gravitational water :** This is the rain water, that percolates deep into the soil layers due to the influence of gravity. Usually gravitational water is not available to the plants as it is beyond the reach of roots. When rain showers follow frequently roots can absorb gravitational water. In water logged soils, gravitational water remains in the upper layers of the soil.

(ii) **Capillary water :** The bulk of the water that remains in the reach of plants after the gravitational Water has drained down is called capillary water. Capillary water remains as thin film in between the soil micelles, and is the main source of water for abosrption by plants.

(iii) **Hygroscopic water :** This represents the water held tightly over the surface of soil micelles due to *adsorption.* Due to this adsorption, hygroscopic water is generally non available to plants.

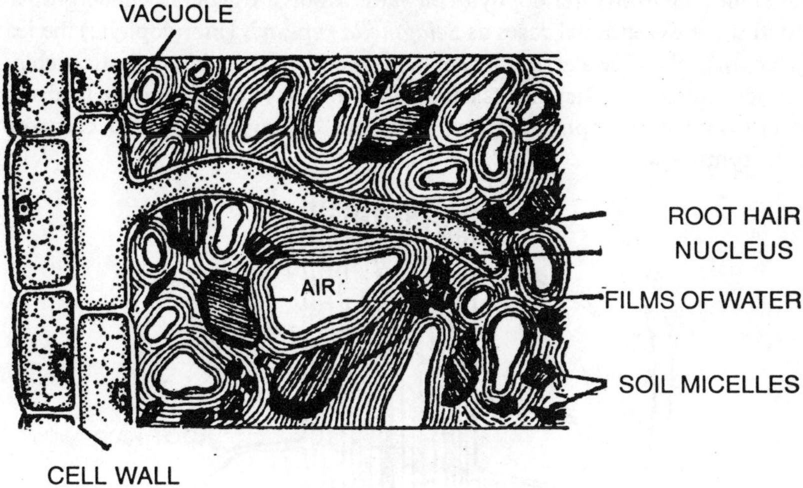

Fig 5.1 Absorption of Water
Soil particles and root hair

(iv) **Run off water :** The water that is present on the surface of the water and is flowing is called run off water. Run off water is usually encountered immediately after a spell of rain.

(v) **Chemical water :** Also called combined water, this represents, water combined with mineral elements of the soil. Plants can not avail this water.

Water absorbing system in plants : Different groups of plants have different methods of abosrbing water from the environment. In lower plants like algae, fungi etc, many of which are aquatic, there is no special organ to absorb water. Water is abosrbed by the general surface of the plants as they are present in the medium of water. In Bryophytes, which are partially terrestrial, separate organs meant to absorb water have been evolved. These are the rhizoids. Rhizoids are unicellular organs, hair like in appearance. In many of the liverworts *(Riccia, Marchantia* etc), two types of rhizoids are seen - *simple.and pegged.* The simple rhizoids have a smooth inner wall, while the pegged rhizoids, have their inner walls infolded. The latter type of rhizoids form an anastamose on the ventral surface of the thallus assuring uniform distribution of water.

In terrestrial plants, soil is the only source of water. As all the parts of the plant body are not in direct contact with the soil, special organs (roots) and special tissues (vasculature) .evolve to provide anchorage, absorption and transportation of water to all the parts of the plant body. Roots are seen in the plant kingdom from ptreridophytes onwards. Roots are generally underground. But in some exceptional cases as *Selaginella rupestris* (Pteridophyte) the leaf surface may absorb water. In some of the flowering plants like epiphytic orchids, the roots are aerial. These roots have a special tissue called *velamen* to absorb water from the atmosphere. These roots are chlorophyllous and carry out photosynthesis.

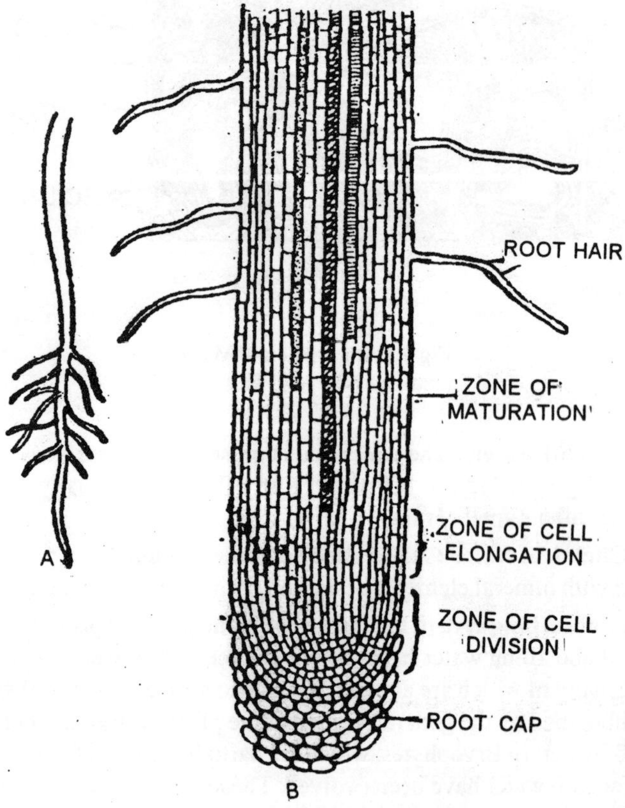

Fig 5.2 Absorption of Water
Structure of a root tip, **A.** Root tip, **B.** Long sectional view of a root tip

In a large majority of plants the roots are underground. Apart from the conventional classification of the root system into tap root system and adventitious root system seen characteristically in dicotyledons and monocotyledons, it (root) may be classified into two categories based on their absorption of water in the soil. These are *deep feeders* and *surface feeders.* In the former, the root system grows very deep in search of water, while in the latter, the root system is not very deep, but spreads very wide a little below the surface of the soil. Both the types of root system are advantageous to the plants depending on the availability of water in soil. In conditions when the surface temp., is high as in deserts, the rain water may get evaporated or it may percolate the lower reaches of the soil very quickly due to loose particles of sand. Insuch a situation, a spreading root system can catch as much amount of surface water as possible before it becomes unavialable. In situations, where water is available, but deep within the soil, deep feeders are of advantage.

The root system is branched and the ultimate branches are called root hairs. The root hairs are unicellular, delicate and vary in length from 1-10 mm. The root hairs are extensions of the epidermal cells. The root hairs are present a little above the region of the root tip where they are present in large numbesrs. This region is called the *root hair zone.* The root hairs remain alive and functional from few days to several months. According to Kramer (1959), maximum water absorbtion takes place in the root hair zone, while the epidermal cells at the root tip, where there is no suberization or cutinization might also absorb some quantity of water.

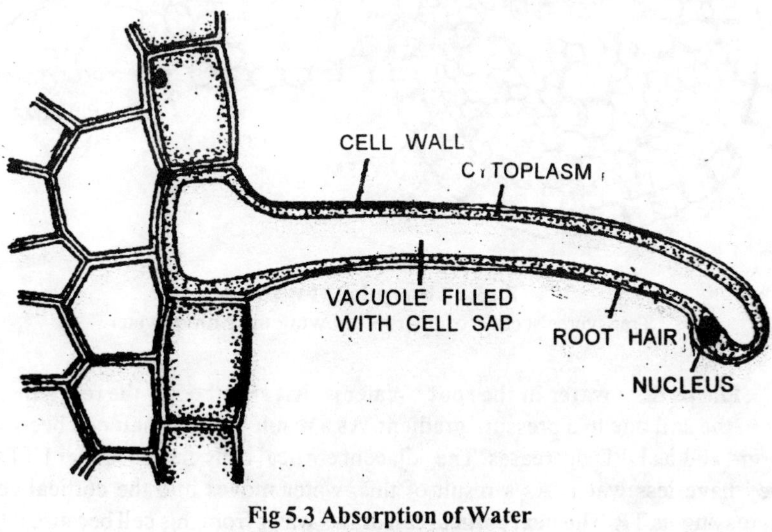

Fig 5.3 Absorption of Water
Structure of root hair

Structure of root hair : A root hair is an elongation of the epidermal cell. It extends at right angles to the piliferous layer. The root hair is 1to 10 mm in length and 10 μ in diameter. The number of root hairs per unit area is a measure of the absorptive capacity of the root. The cell wall of root hair has two layers.

The outer layer is mainly composed of pectic substances, while the inner layer is made up of cellulose. The wall composition varies in some instances as in *Taxus baccata,* where the outer wall is gelatinized and soil particles get integrated with it. In Corn, the outer cell wall of root hair has no pectin, while in *Tradescantia,* callose is present at the tip of the root hairs. Inside the cell is a large vacuole, which is continuous with the vacuole of the epidermal cell. The vacole is filled with cell sap. Between the vacuole and the inner cell wall is a thin layer of cytoplasm, which is also continuous with that of the epidermal cell. There is no separate nucleus in the root hair. The nucleus of the epidermal cell controls the activity of the root hair also. The cell wall of root hair is permeable to water. The cytoplasmic layer functions as a differentially permeable membrane.

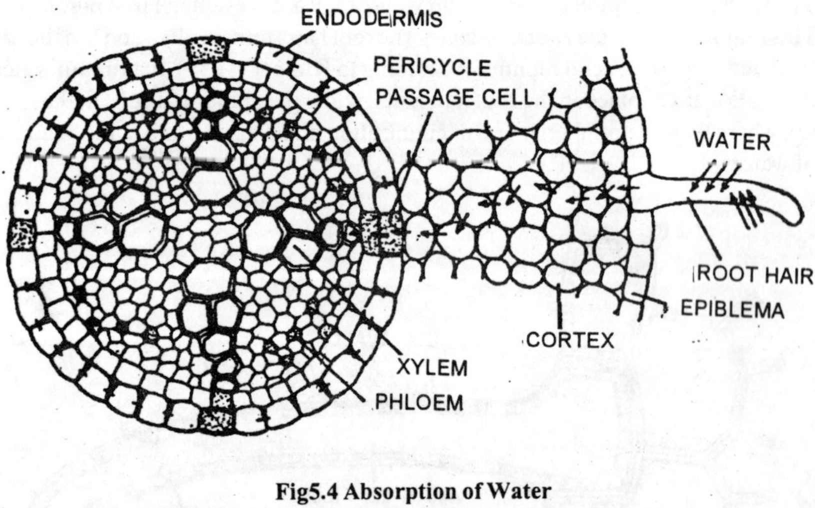

Fig5.4 Absorption of Water
Transverse section of the root showing the path of water

The path of water in the root : Water is first absorbed by the root hair cell from the soil due to a pressure gradient. As a result, the root hair cell becomes turgid and its DPD decreases. The adjacent cortical cells have a higher DPD as they have less water. As a result of this, water moves into the cortical cells increasing its TP. The next cortical cell draws water from this cell because it has less TP. In this way, through a series of concentration gradient water flows from one cortical cell into the other and finally reaches the endodermis. According to Sutcliffle (1969), the movement of the water through cortex is in response to a water potential gradient.

There are two ways of radial movement of water (i) Cell to cell movement across the root and (ii) movement of water along the water filled spaces in the cell wall forming a continuous system from soil water up to the endodermis. The flow of water meets with an impregnable barrier in endodermis as its cells have thick strips on the radial and tangential walls called *Casparion strips*. However there are certain thin walled cells in between the endodermal cells called *passage cells*. It is through these cells, water enters into the stele. The passage cells are usually positioned opposite the protoxylem elements. The passage cells allow the entry of water into the cells of the pericycle, which in turn become turgid. The protoxylem elements situated next to the pericycle always have a higher DPD and they absorb water from the cells of the pericycle. There are various views regarding the unidirectional flow of water into the xylem elements.

According to Lepeschkin, the parenchyma cell next to the xylem element has differential permeability at different sides. The side towards the xylem element is more permeable allowing the entry of water into it.

According to Priestley, the xylem elements always have a higher DPD due to transpiration and hence draw water from the surrounding cells.

Factors affecting absorption of water : There are a number of internal and external factors that affect the rate of water absorption. These are

(a) **Available soil water :** Of the different types of soil water it is only capillary water that is available to the roots. Usually absorption of water is uniform in plants, when soil water is in between the field capacity and permanent wilting percentage. When the quantity of water is either above the field capacity or below the wilting percentage absorption is affected. Increased water decreases soil aeration and thus affects absorption. High water contents may even bring about wilting.

(b) **Temperature of soil :** Alterations in soil temperature beyond the tolerance range of roots affects the abosrption of water. A range between 20^{0}C - 30^{0}C is most suitable for water absorption. A low temperature generally retards the rate of absorption because

(a) roots elongate slowly

(b) metabolic activities like respiration will be low

(c) rate of diffusion comes down

(d) viscocity of protoplasm increases and it becomes less permeable to water.

A moderate increase of temperature will enhance the rate of water absorption, while a very high temperature kills the living tissues of the root.

(c) **Soil aeration :** Proper aeration of the soil is necessary for normal rate of water absorption. The amount of soil air depends on the type of soil; while clay soils have very little soil air, sandy soils have too much of it. In medium soils, there is adquate soil air. Proper soil aeration is necessary for roots, both physically and physiologically. In well aerated soils, the soil particles are loose and root hairs can easily penetrate the soil. Further, adequate amount of soil air is necessary for respiration of roots. Roots fail to grow in poorly aerated soils. Poorly aerated soils tend to accumulate CO_2 released by root cells due to respiration. This accumulation increases the viscosity of protoplasm and reduces its permeability. In certain plants like *Oryza, Typha* etc., there is genetic adaptation to grow in water logged soils.

(d) **Concentration of soil solution :** The soil water is not present in its pure state. It is usually combined with the mineral elements of the soil to form soil solution. This concentration of the soil solution is a vital factor in the diffusion of water into the root hair cells. A very high concentration of soil solution (as in salt marshes) will in fact attract water from the roots than *vice versa.* This is the reason, why many of the plants growing in marshes exhibit xerophytic characters (physiological dryness).

(e) **Transpiration of the plant :** The rate of absorption of water is approximately equal to that of transpiration. A high rate of transpiration will directly influence absorption of water due to cohesion among water molecules.

(f) **Types of root system :** The efficiency of the root system greatly influences the rate of absorption of water. In general, an extensive branching in the root system increases the absorbing regions *ie* root hairs. But more important than the branching is the number of root hairs per unit area of the root tip. In coniferous plants, there are no root hairs, but the mycorrhizal hyphae in association with the roots help in the absorption of water.

(g) **Metabolism of the plant :** A proper metabolic rate is very essential for water absorption. Poor metabolic rate reduces the quantum of available energy for all the physiological activities including absorption of water. Factors that inhibit respiration are also known to affect absorption of water, even though it is open to question whether respiratory energy is directly involved.

MECHANISM OF ABSORPTION

Mainly there are two ways in which water is absorbed by the roots. These are passive absorption and active absorption.

Passive absorption : The roots do not play any role in the absorption of water in the sense, the force or forces responsible for water absorption are generated elsewhere in the plant body. It is widely beleived that transpiration has a direct bearing on the rate of water absorption. That is why in rapidly transpiring plants, rate of absorption *of* water is very high. In a rapidly transpiring plant, a continuous gradient of DPD is built up due to evaporation of water. The mesophyll cells haye a higher DPD, and as a result, water in the xylem elements are attracted towards them. This creates a tension (DPD) in the xylem elements which is transmitted down due to the cohesion of water molecules. The tension or DPD reaching the lower tip of the xylem is carried across the root cortical cells and ultimately to the root hair cell. As a result of this continuous chain of DPD, water migrates radially across the root cortex and travelling through the xylem finally reaches the leaf mesophyll cells, from where it escapes. Hence transpiration force or pull or suction force generated away from the roots, is responsible for the absorption of water by the roots. It would be more

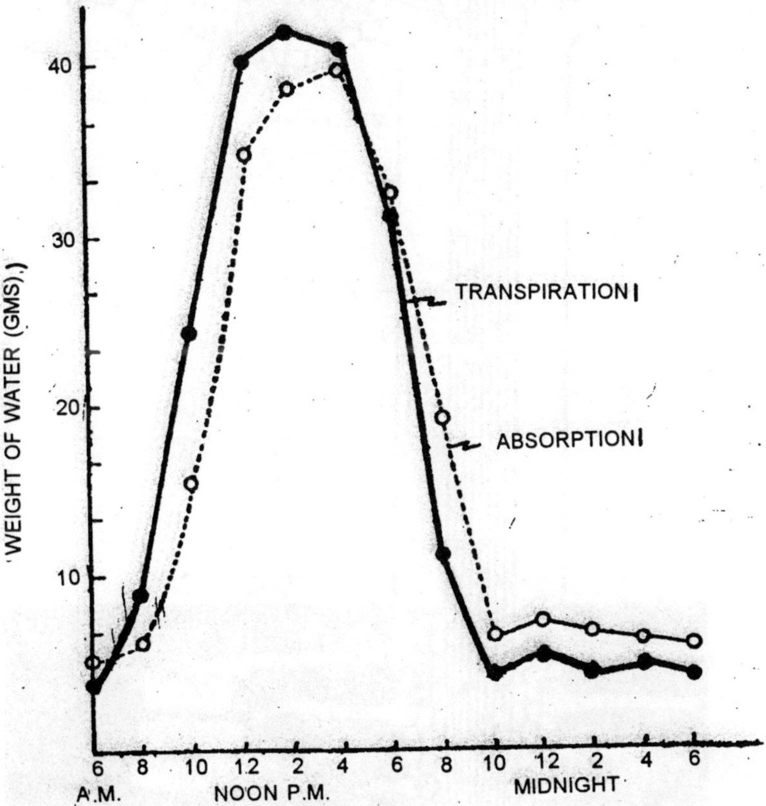

Fig 5.5 Absorption of Water

Graph showing the direct relationship between transpiration and passive absorption appropriate to use the term absorption of water *through the rots* rather than *by the roots* because roots are passive and do not influence the water uptake in anyway. No event taking place in the root will influence the absorption of water (either way).

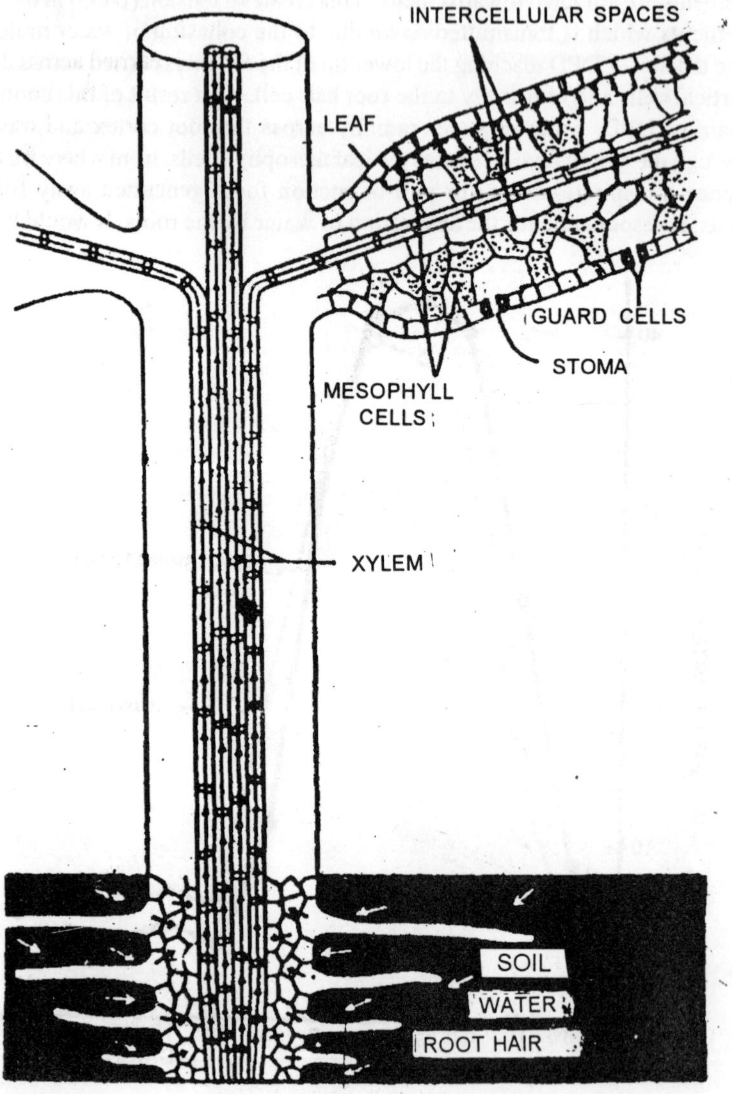

Fig 5.6 Absorption of Water
Diagrammatic representation of passive absorption of water

When the plants are rapidly transpiring, it has been estimated, that the DPD in root hair cells increases by -3 to -5 bars, whereas the DPD of the soil solution is less than - 1 bar. It is this higher DPD that forces large quantity of water into the roots, ultimately to be lifted up to the leaves. Evidence for passive absorption is obtained from the fact, that the rate of transpiration is approximately equal to absorption of water, and any inhibition of transpiration inhibits water absorption also (Kramer 1937, Lachenmeir, 1932). To sum up, it may be said that in passive absorption *roots provide merely a conduit for the entry of water in the plants.*

Active absorption : The mechanism of water absorption due to the forces generated in the root itself is called active absorption. According to this theory, roots are not just passive surfaces, witnessing the passsage of water through them, but actually play a dominant role in governing the quantum and the rate of entry of water. The proponents of active absorption beleive that all the events leading to the intake of water into the root cells are controlled by the root surfaces. There seems to be some controversy regarding the actual purport of the term *active,* whether its use should be restricted to only in instances where metabolic energy is involved, or whether the term can be used in the sense of roots taking an active role in absorption. Hence in order to avoid confusion the term active should be qualified with a prefix indicating the specific source of energy. Accordingly active absorption is of the following two categories.

(i) Osmotic active absorption and
(ii) Non osmotic active absorption.

Osmotic active absorption : Atkins (1961) and Priestley (1922) who postulated an osmotic theory for water absorption, opined that, the difference in the concentration gradient between cell sap and soil solution is responsible for the entry of water into the roots. According to them, the situation in the soil near the root presents an ideal osmotic system. There are two solvent systems (soil solution and cell sap) differing in concentration and separated by means of a selectively permeable barrier (cell membrane). As the water potential of cell sap has a higher negative value than the soil solution osmotic migration of the solvent takes place into the cell.

Investigations have revealed that the cell sap OP may be about 2 atm and may go up to 8 atm in some circumstances. The OP of soil solution in most cases is less than 1 atm. As a result there is endosmosis.

A number of plant physiologists have supported the osmotic absorption, by demonstrating a close link between the phenomenon of sap exudation (root pressure - see ascent of sap) and water absorption. Some of these scientists are Lundegardh (1950), Broyer (1951), Hoagland (1944) etc. When an actively water absorbing plant is decapitated, sap exudes out of the cut end of the stem indicating active absorption of water. According to Atkins, the xylem sap always has a high concentration of solutes due to the release of sugars into them by the adjoining parenchyma cells. Anderson and House (1967) have reported that salts are absorbed by the xylem elements containing living protoplasts. Priestly (1922) on the other hand opined that a cell differentiating into xylem gets higher salt concentration because of its contents.

Objections to osmotic theory : The main objection comes from the fact, that the sap concentration in xylem is not always high, as a result a gradient may not exist between cortical cells and xylem. (Xylem sap has 2 atom; cortical cell have 3-4). Secondly root pressure is not universal in plants.

Non osmotic active absorption s: Many physiologists beleive that water absorption is driven through the expenditure of metabolic (respiratory) energy. Absorption of water has been observed against the concentration gradient and even when there is a gradient, the rate of absorption far exceeds what is possible due to a concentration gradient (Bennet Clark 1936, Thimann 1951, Bogen and Prell, 1953 etc). The following are the evidences to suggest, the involvement of respiratory energy in water absorption.

(a) Many of the respiratory poisons (Malonate etc) also inhibit water absorption.

(b) KCN which reduces metabolic activities also reduces water absorption.

(c) Respiratory promoters increase the rate of absorption (e.g. hormones)

(d) Low temperature, which retards the rate of energy release during repsiration also, reduces water intake.

(e) Reinders (1942), Kelley (1947) etc., have discovered that aerobic conditions are necessary for water intake.

(f) Wilting of roots occurs in water logged soils as they have poor aeration.

(g) Only living cells in roots absorb water (?)

(h) Bennett Clark et al (1936) showed that OP of cell sap determined by plasmolytic method yields a more negative value than the cryoscopic method. According to them, this indicates a non osmotic component in water absorption. They suggested a secretion force produced at the expense of metabolic energy, pumping water into the vacuole. Kramer (1956) on the other hand opines that the discrepancy in OP is due to experimental errors.

(i) House and Findlay (1966) found that the OP of xylem sap was equal to that of external solution and hence opined the involvement of a respiratory energy in water intake.

(j) A diurnal periodicity has been observed in water absorption indicating fluctuations in the available energy from metabolic processes.

Many objections and doubts have been raised by workers for the active uptake of water using respiratory energy. Some of these are listed below.

(i) Low temperature, and lack of O_2 apparently increase the viscosity of cytoplasm and reduce the permeability. This does not indicate the involvement of respiratory energy. Lopushinsky (1964) argued that lack of CO_2 inhibits water absorption by passive mechanism also, brought about by concentration gradient.

(ii) There is no irrefutable evidence to directly link water absorption and respiration. Respiratory poisons by blocking the release of energy, generally upset all the physiological systems. Naturally a weakened cell cannot carry on water absorption.

(iii) Cytoplasm is more permeable to water than to solutes and hence a great amount of energy is needed to maintain the positive gradient for water than for solutes. According to Levitt (1947), in the sugar beet plant, the whole of the energy released by a cell is not enough to maintain water potential of more than one atm.

(iv) Auxin induced water uptake, cannot be linked to increased availability of respiratory energy. Auxins are known to reduce TP and increase the permeability of the membranes.

In view of this, Kozoiwski (1964) opined that the bulk of water intake in plants is by the passive mechanism and active up take if any is negligible. Steward (1959) also argued that "the existence of water deficit in root cells will actually hinder or even prevent active absorption".

Electro osmotic theory : According to Keller (1930), water is taken in by the xylem elements due to *electroosmosis.* The membranes in contact with water are negatively charged and water moves to the interior of the cells due to this charge. Fenson (1958) beleived in an *electrokinetic* absorption of water. But the observed differences in the electrical potential are too meagre to bring about water absorption in large quantities (Lundepardh 1940 Blinks 1951).

6

ASCENT OF SAP

In terrestrial plants whether they are prostrate plants, or herbs of about 1 or 2 feet high or tall trees reaching a height of more than 350 feet, the only source of water for all the cells, whether it is a cell of the root or a cell of the apical bud (sitting pretty on a pedestal of 350 feet high), is soil. The herb or a tree absorbs water from the soil and after reaching the lower reaches of xylem in the root, it (water) travels up. In the case of small herbs the distance to be travelled by water to reach the top is not much; it may be 2-3 feet at the maximum. But if we consider the situation in trees, where even the apical bud has to be served by water taken from the soil, the water has to travel up a considerable distance. In other words, water has to be lifted up to a height of 300 - 350 feet. Indeed an enormous amount of energy is required to lift the water to such dizzy heights, and that too against the force of gravity. What type of energy is involved in this task? Does it involve biological energy or is it merely physical? Conclusive answers to these questions are still wanting. In this chapter we shall study the various theories that account for the upward movement of water (ascent of sap).

Path of ascent of sap : There is no doubt that, the water travels up along the vascular tissue, as it (vasculature) is ideally suited, both in structure and location to perform this task. But the vascular tissue has two components viz. xylem and phloem. Between these two which tissue is the path for translocation of water? Is it only xylem or phloem or both? The following experiment conducted with the balsam plant will show that xylem is the only element through which water moves up. The balsam plant has a transparent stem so that the interior is visible.

Cut a few branches of the balsam plant and place them in a beaker containing coloured water. After some time coloured streaks appear in the transparent stems indicating the course of water. Cut a transverse section of the stem and observe it under the microscope; only the xylem elements are coloured indicating that they constitute the path for ascent of sap. Phloem is not coloured indicating it has no role in this. The fact that only xylem is the path for ascent of sap may further be confirmed by ringing experiment. In ringing experiment, a girdle of bark external to xylem is removed and it has been observed that this has no effect on ascent of sap.

Another point that needs confirmation in the path of ascent of sap is, whether the water moves in the xylem along the lumen or along the cell wall. Because, the cell wall of xylem having hydrophilic colloids can imbibe water and trasport it to upper regions. This can be ascertained by dipping the balsam twing in molten paraffin for some time and then placing it in coloured water. On dipping in paraffin, the lumen of the xylem element gets blocked. No streaks appear in such a twig indicating that there is no upward movement of water, because of thc blockade of lumen by paraffin. Hence it can be ascertained that water moves up in the plant along the lumen of xylem elements.

MECHANISM OF ASCENT OF SAP

The problem of the forces involved in ascent of sap has attracted the attention of plant physiologists, since a long time. A number of theories have been proposed from time to time. All these can be grouped under three headings, a) Vital force theories, b) Root pressure theory and c) Physical force theories.

VITAL THEORIES

There are various opinions in the vital theories of ascent of sap as to the exact source of energy required for the movement of water. But all the proponents of vital theories are unanimous in that, the xylem, though it happens to be the path of ascent of sap does not have a role in the process. Instead, the living cells of the stem, with their vital energy play a dominant part in the same. Some of the proponents of vital theories are - Westermeir (1883 - 1884), Godlewski (1884), Janse (1887), Jagdish Chandra Bose (1932), Mollisch (1928-1929) etc.

According tho Westermeir (1884), the xylem elements only act as water reservoirs, while the actual force for ascent of sap is produced in the wood parenchyma. Godlewski (1884) another proponent of vital theory proposed his relay pump hypothesis to account for the upward movement of water. According to him the parenchyma cells surrounding the xylem element are the ones that are responsible for ascent of sap. He opined that there is a rhythmic change of osmotic pressure in the vasculature between the vessels, parenchyma, and medullary rays.

As a result of this rhythmic action, water is pumped up in the plant. In order to explain the mechanism of this relay pump hypothesis, we have to visualize an alternate stepwise arrangement of xylem and phloem. When the OP of the surrounding parenchyma cell increases, it draws water from the cortical cells which have a higher TP. The absorbed water lowers the OP of the parenchyma cell. This cell which is in contact with a xylem element empties its water to the latter (xylem), because it has a higher OP. The xylem element now will have less OP than the parenchyma cell situated above, as a result water is drawn into the parenchyma cell. In this way water pumped into the xylem by parenchyma and again drawn back into another parenchym cell because of stepwise OP. Hence water moves alternately into xylem - parenchyma - xylem. This theory is called the relay pump, because water is relayed between parenchyma and xylem. Janse (1887), supported this hypothesis and showed that if the lower portion of a branch is killed, .the leaves above are affected. Many evidences for this theory were provided by Ursprug (1907), Ewart (1908), Schwendener (1893) etc. But none of the evidence could stand the scientific scrutiny of Strassburger, who showed that, ascent of sap was independent of the living cells in the plant. He showed that long after the killing of cells surrounding xylem, water continued to rise in the xylem elements. Experiments of Overton (1911) and MacDougal (1929) confirmed the findings of Strassburger.

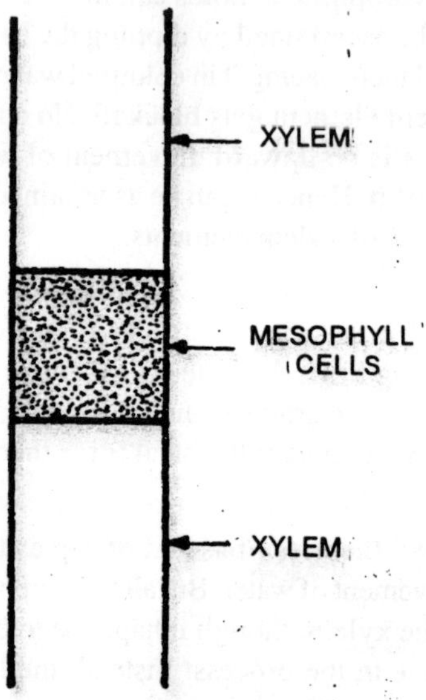

Fig 6.1 Ascent of Sap
Hypothetical arrangement of xylem and parenchyma

Another serious objection to relay pump hypothesis comes from the structural arrangement of xylem and parenchyma. In order to be efficient in pumping water into the xylem as visualised in realy pump hypothesis parenchyma cells should be situated in between the xylem elements and not at their sides as they are seen in the plant. The work of Kutzmann (1966) also has shown, that injecting metabolic inhibitors (picric acid, mercuric chloride etc) into trees seems to have no impact on ascent of sap.

Jagdish Chandra Bose (1923) an eminent physicist and a plant physiologist strongly supported the vital theory of ascent of sap. He was of the opinion that the innermost layer of cortical cells abutting the endodermis are of a special category as they are endodwed with rhythmic pulsations, which is electrical in nature. These cells 'are responsible for the pumping up of water into the xylem.

Bose's theory, called *pulsation theory* envisages that the alternating contraction and expansion of these cells forces water into the xylem. In order to prove the electrical pulsation of the innermost cortical cells, Bose invented an instrument called the *Cresograph,* which consisted of an electric probe connected to a galvanometre. When the fine needle of the probe was pierced into the cortex there were minor oscillations. When the probe reached the inner cortical cells however, the oscillations in the galvanometre indicator became violent and continuous. This showed the high electrical activity of the inner cortical cells. Bose beleived that these cells expand when they absorbed water from the outer cortical cells and after pushing water into the xylem they contracted. Each contraction and expansion took 14 secs to several minutes.

Molisch (1929), who supported Bose's theory, experimentally showed, the electrical activity of the inner layer of cortex could be increased when treated with certain chemicals.

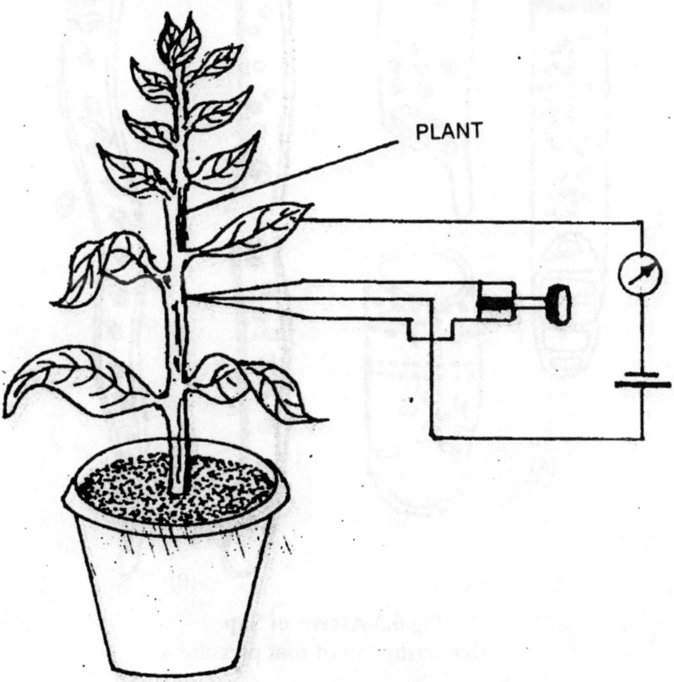

Fig 6.2 Ascent of Sap
Cresograph of J.C. Bose

Shull, MacDougal and others have shown that the pulsatory activity of the inner cortical cells has no relation to ascent of sap. Further, in order to account for the rate of ascent of sap, the pulsation rate must be far higher than what is demonstrated by Bose. Benedict (1927) showed that the rate of ascent of water, is 8,000 - 30,000 times higher than the pulsation rate recorded by Bose.

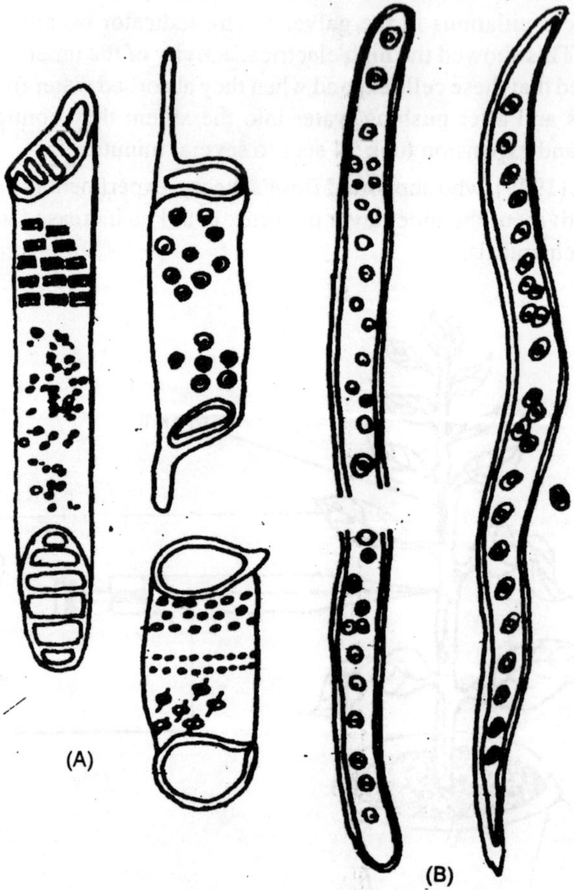

Fig 6.3 Ascent of Sap
Demonstration of root pressure

The claim of some plant physiologists that atleast in aquatic plants, ascent of sap is due to vital activities of the stem, has been refuted by Wilson (1947). Who argued that uptake of water by aquatic stems is caused by the readjustment of the water balance.

The experiment of Strassburger (1891) with the oak tree is worth mentioning. A 75 year old tree was cut at its base and kept in a solution of.picric acid which is poisonous to cells. Picric acid slowly ascended" to the top and killed all the living cells on its way. Basic fuchsin dye was added to the picric acid solution after three days, and the dye reached to the top of the cells, even though there were no living cells on the way.

ROOT PRESSURE THEORY

If a well watered plant is cut a few inches above the ground level, a sap exudes out with some force. This is called *sap exudation* or *bleeding* and is found commonly in most of the plants. Joseph Priestly was the first one to have observed this phenomenon. According to him a sort of hydrostatic pressure builds up in the roots due to continuous absorption of water, and this is responsible for the upward flow of water. Stephen Hales (1727), who also observed this, coined the term *root pressure* to explain this phenomenon. He fixed a tube of 8 mm diametre to the cut end of the stump and noticed that in a few days time, water can rise up to 6 metres. This is due to the pressure built up by the accumulated water. The water absorbed by the roots reaches endodermis, after travelling across cortex, where it meets with resistance for its smooth flow. As water continues to be absorbed, a hydrostatic pressure is built up, which forces water into the xylem resulting in its upward flow. This is called root pressure. Root pressure may be defined as the *hydrostatic pressure built up by water in the outer boundary of the stele due to continuous absorption.* Stephen Hales (1827) recorded a pressure of over one atm in the roots. However it has been observed that root pressure is largely an osmotic phenomenon and disappears, when isotonic or hypertonic solutions come into contact with roots, when there is a possibility of no gradient or reverse gradient away from the interior of the root. Root pressure may be demonstrated by attaching a mercury manometre to cut end of the stem. Within a short time, the level of mercury rises up indicating upward movement of sap due to root pressure.

Respiratory poisons and depleted oxygen supply are known to affect root pressure indicating the inhibition of the supply of biological energy. This shows that root pressure is a phenomenon concerned with the metabolic activities of the living cell. Inhibition of carbohydrate metabolism and obstruction to mineral salt absorption are also known to affect root pressure. These factors largely affect energy supply through respiration rather than affecting the permeability of protoplasm. In view of this. Stocking (1956) argues that root pressure is an active process and defines it *"a pressure developing in the tracheary elements of the xylem as a result of the metabolic activities of the roots"*.

The phenomenon of root pressure and the resultant upward movement of water in xylem made many physiologists to opine, that ascent of sap is largely due to root pressure. But the following objections have been raised against root pressure being responsible for ascent of sap.

1. Root pressure is not a universal phenomenon, but ascent of sap is. In Gymnosperms, which includes some of the tallest plants, with a high rate of ascent of water, little or no root pressure is observed.
2. There is no relationship between root pressure and ascent of sap in terms of rate of the process. During summer, when the demand for water is more and the rate of ascent of sap is high, a very low root pressure has been recorded in plants. On the other hand in autumn, when the rate of ascent of sap is low, a high root pressure has been recorded.
3. Ascent of sap continues even in the absence of roots.
4. The magnitude of root pressure is not sufficient to account for ascent of sap in such tall trees as *Eucalyptus* and *Sequoia.*
5. In plants which are actively transpiring, and require a regular supply of water to the distal regions, very low root pressure has been recorded. On the other hand, when the rate of transpiration is low a high root pressure has been observed.
6. The quantity of water that exudes out due to root pressure constitutes barely 5% of the amount lost due to transpiration. (The main source of water supply for transpiration is ascent of sap).
7. The maximum root pressure seen under most favourable conditions is about 2 atm. This can account for water rise of only up to 60- 62 feet. Hence this cannot account for water rise in plants taller than 60-62 feet.
8. An important objection, that delinks root pressure and ascent of sap is that the latter can take place even in the absence of living cells, while the former is beleived to be an active process linked to the supply of metabolic energy.
9. The xylem sap, under normal conditions will be under tension (high OP) rather than pressure (low OP), indicates that root pressure does not have a significant role in upward movement of water.

Some physiologists however opine that atleast during certain conditions, like interruption in water column, root pressure helps to hydrate the xylem elements.

PHYSICAL FORCE THEORIES

1. Capillary force theory : When a narrow tube is dipped in water, due to surface tension it (water) rises up in the tube up to a certain height. This phenomenon has been observed by every body, who has dipped a straw in a can of soft drink. The similarity in situation between a capillary tube and xylem element made Boehm (1809), to argue that capillary force is responsible for ascent of sap. But there are several objections to this theory.

(a) Capillary force is inversely proportional to the diametre of the tube; larger the diameter, lower will be the height to which water can rise. According to this, herbs should have vessels with very narrow lumen. But this is not true. And also considering the fact, that diameter of the capillary tube must decrease with the increased height to which water has to rise, the xylem must not have any lumen at all in tall trees!

(b) The smallest diameter observed in tracheids is about 0.02 mm. with this diameter, if water were to rise by capillarity, it cannot travel beyond a height of 150 cms. In larger vessels which have a diameter of 0.5 mm water cannot rise beyond a height of 6 cms. only in capillaries with a dia, of 1μ water can travel up to 29.95 metres, but, the rise of water is not through lumen, but through the wall due to imbibitional force. But ascent of sap takes place through the lumen and not through the wall.

(c) In lower terrestrial plants, which are vesselless and have only tracheids, capillarity cannot function due to the presence of cross walls at the end of every cell.

(d) For capillarity, the capillary tube should be directly connected to water source. But this is not the case in the plant. The lower ends of xylem elements are not dipped in soil water.

(e) A free surface must be present in the xylem vessels for maintaining capillarity.

2. Imbibitional theory : The wall of xylem elements have hydrophilic colloids and can imbibe large quantities of water. Hence many physiologists, notably Sachs (1978), Unger (1868), etc., opined that imbibitional force is responsible for ascent of sap. While it is true that imbibitional movement of water is very slow, the pressure (100-1000 atm) is quite sufficient to carry water to any height. But the fact that water moves in the lumen and not through the walls, clearly is a point against imbibition being responsible for ascent of sap.

3. Atmospheric pressure theory : The fact that the xylem is vertically elongated and water rising up as in a tube, made many people think, that atmospheric pressure is responsible for upward movement of water. Water moves up in the xylem vessels to fill up the drop in atmospheric pressure caused due to loss of water during transpiration. There are two main objections to this theory.

(a) For the operation of atmospheric pressure the lower end of the column should be free. But this is not found in plants, xylem elements do not directly open into soil water.

(b) The maximum height that can be reached by water, assuming it to be due to atmospheric pressure is not more than 30 feet. Further, there is no vacuum at the upper ends of the plant which is necessary for atmospheric pressure to elevate a column of liquid.

4. Jamin's chain theory" According to Jamin, water and air are alternately arranged in the plant. When air expands it carries the water up along the xylem elements. This theory cannot be accepted because, there is no evidence of alternate arrangement of water and air in the plant. Further the rate of ascent of sap cannot be accounted by the force of the expanding air within a plant, assuming it to be present.

5. Transpiration pull or water cohesion theory : This theory which has been widely accepted inspite of some short comings, was first proposed by Dixon and Joly in 1894 and Askenasy (1895). The theory was proposed again (to answer the objections raised against it) with experimental detail with the support of several plant physiologists like Renner (1911), Curtis and Clark (1951), Bonner and Galston (1952), Kramer and Koziowskji (1960), Milbum and Johnson (1966), Hammel (1967) and Levitt (1969). There are three main aspects in this theory. These are -

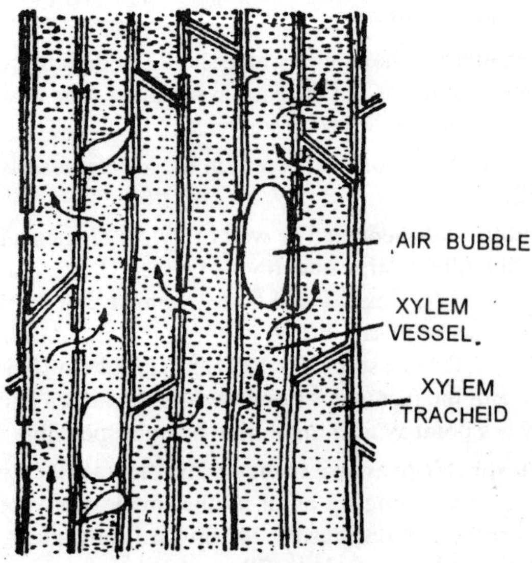

Fig 6.4 Ascent of Sap

Diagram showing the continuity of water system within the xylem

(a) Cohesion of water molecules : Attraction and adhesion to one another is called cohesion. Water molecules have a strong cohesive force due to which they are held tightly to one another and cannot be easil) separated. In addition to this, water molecules have an adhesive property also which makes them have a strong attraction to the inner wall of xylem.

Hence water molecules cannot be easily torn away from the wall of the xylem. Some calculations have been made about the strength of cohesion and many physiologists agree that it is in the range of 350 atm far in excess of what is required for ascent of sap. However there are certain forces within the plant itself that work against the cohesive force of water these are-a) the weight of the column, b) resistance encountered by water when it has to cross tissues, c) resistance by the conducting ducts. In addition to this, there is the downward pull of the gravity on the water column. All these add up to about 50 atm. But as has been pointed out already, the observed cohesive force far exceeds thc force that can brcak it. In order to answer some objections raised against the maintainability of water column, Dixon (1914) measured the cohesive force of water in the xylem sap of *Ilex aquifolium.* He took a J shaped glass tube and partially filled it with xylem sap. The sap was heated untill it almost filled the lone arm. The tube was then sealed and kept inclined, so that the long arm was completely filled by the sap. Afterwards the tube was inverted to see, whether the water column would break in the long arm. The water column did not break due to the cohesive and adhesive force among its molecules. The water column would break, on cooling. Dixon tested several samples of the sap and observed the extent of cooling required to break the column. From this he calculated, that the sap had a cohesive strength of 133-207 atm. The works of Gremidge (1957), Loomis (1960) etc., have supported the existence of cohesive force among water molecules, but not as high as reported by Dixon. They opine that the maximum cohesive strength is about 30 atm. Considering the fact that even a force of one atm can maintain a height of 30 feet, 30 atm pressure is more than adequate even for the tallest trees. Theorcticaly 30 atms can account for about 900 feet. The water column can be streched also without breaking. According to Fischer, a force of 13,000 bars is required to break the water column.

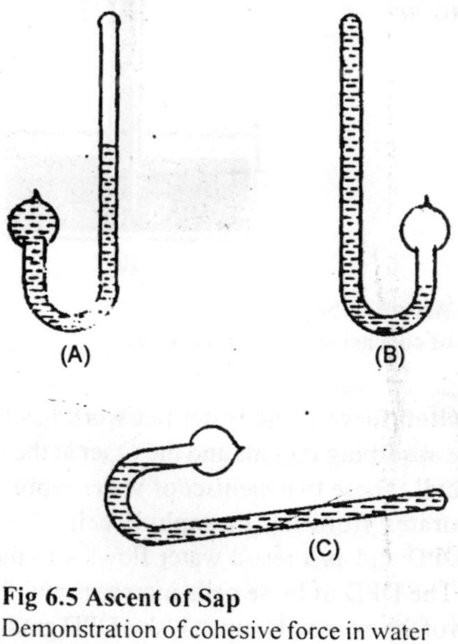

Fig 6.5 Ascent of Sap
Demonstration of cohesive force in water (Dixon 1914)

(b) Continuity of water column within tlie plant : According to Dixon and Joly, the entire water column in the plant, starting from root hair cells, cortical cells, xylem, to mesophyll cells of the leaf around a stoma is a continuous chain, so that any influence in the upper part or the lower part is felt by the entire column. The water columns present in different xylem vessels are not independent. They are laterally connected. In other words, the whole of the water present in the plant from root tip to stem tip is a single continuous system.

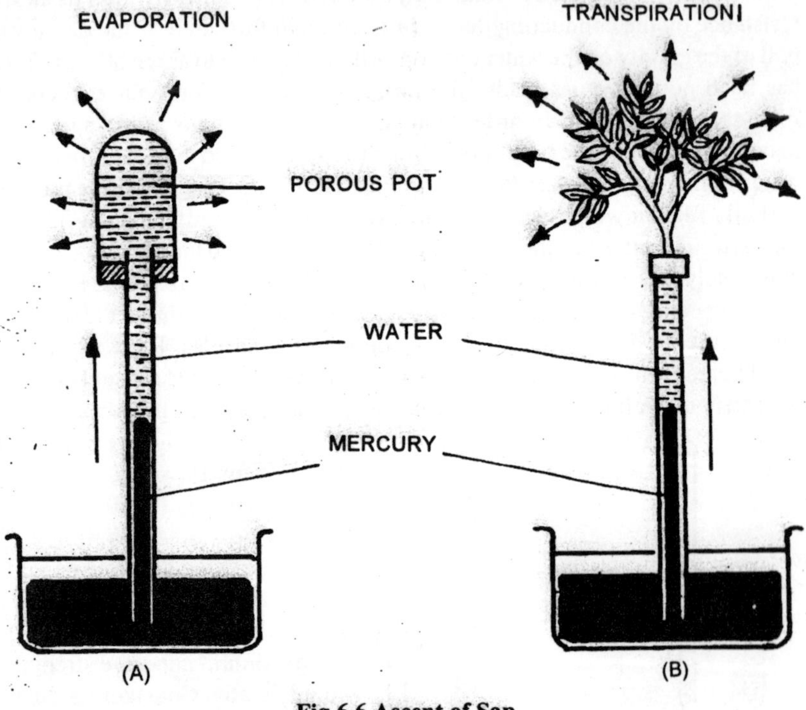

Fig 6.6 Ascent of Sap
Physical demonstration of cohension - tension of water

(c) Transpiration pull or suction force : The water net work has two terminals, one at the root tip near the absorbing regions and the other at the sub stomatal cavities in the leaf mesophyll. These two *menisci* of water represent the two extremities. Water evaporates from the mesophyll cells due to transpiration. This increases their DPD and as a result water flows into them from the adjacent mesophyll cells. The DPD of these cells increases and they attract water from the mesophyll cells of the interior. Ultimately this DPD gradient reaches the mesophyll cells abutting the vasculature, which (mesophyll cells) draw water from the xylem elements. The DPD of water in xylem increases as a result of continued drawal of water by the mesophyll cells. Continued transpiration in the leaf creates a tension (negative pressure) in the xylem elements in the upper parts. Since the water column is continuous and unbroken, this tension is transported down into the root up to the area of absorption and water moves up due to this force. In other words water is pulled up due to the suction force generated by transpiration. This may be likened to the pulling up of rope (water column) through a tube (xylem).

Some calculations have been made about the amount of tension (negative pressure) created in the xylem due to transpiration. Since 1 atm is equivalent to a height of 30 feet, in tall trees of more than 350 feet high, at least 11-12 atm is required to lift the water up by means of suction force. In many of the tall plants a negative pressure of up to 20 atms have been observed.

Timmerman (1965) has calculated that a gradient of 0.15 bar per metre is required for the upward movement of water. At this rate, some of the tallest trees require a bar potential of 40 to lift the water up. The tension in the xylem sap can not be measured directly. Indirect measurements have shown as mcuh as -77 bars (Slatyer 1957). Arcichovski and Ossipov (1931), two Russian scientists have reported a pressure of -142.9 bars in some plants.

Direct measurement of tension in the xylem sap was first done by Scholander and associates (1965) using a **pressure bomb** method. They introduced a freshly cut twig into the pressure bomb appartus with the cut end projecting out through an air tight seal. They applied counter pressure on the twig (inside the pressure bomb) by means of nitrogen gas to make the water travel back and come out of the cut end. This required a force of approximately -80 bars, indicating the existence of an equivalent force to move the water up in the shoot.

Evidences for transpiration pull theory

1. Measurement of tension in xylem sap indicates adequate amount of pressure to lift the water up even in tallest trees.
2. It is a physical process and does not involve metabolic energy.
3. The tensile strength of xylem sap is in the region of 25-300 atm which is sufficient to withstand any force that attempts to break it.
4. Kramer and Koziowski (1960) made dendrographic measurements of tree trunks and showed a diurnal variation of the girth. There is decrease during day time as a result of shrinkage of cells. This shrinkage is due to water loss and the resultant tension in the xylem elements.

Objections : One of the serious objections raised against the continuity of water column is in view of several physical barriers in the form of tissues. In addition, a number of air bubbles present in the xylem sap disturb the continuity of the water column. Copeland (1902) stated that the air bubbles do not allow continuity of water column. Preston (1958) also expressed the same opinion. There is no doubt that bubbles do get into the xylem sap, but soon get dissolved in water column. Hammel (1967) has pointed out that the air blocks get dissolved in Gymnosperms (growing in high altitudes) during melting of ice.

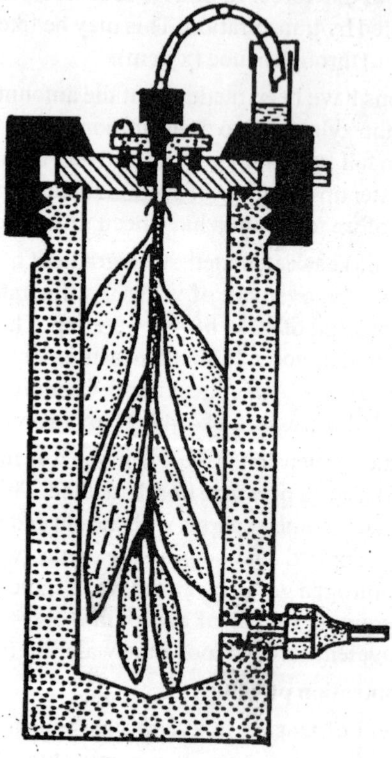

Fig 6.7 Ascent of Sap
Direct measurement of tension in xylem sap using a pressure bomb

Another objection that is raised against the continuity of water column is that trees, whose trunks are cut at sides alternately on right and left still have an unceasing flow of sap. Postlethwait and Rogers (1958) have interpreted this actually as a proof of continuity of water column in the face of disturbances. By using radio active isotopes (P^{32}) they have shown, that water molecules, whenever they face a dead end, avoid it, move internally and maintain the continuity of the water system. Scholander et al (1957), showed that whenever the stem is cut, the cut vessels may not function, but others continue to-do so. Even in the cut vessels, water continued to move above and below the cut, apparently because it was in connection with water of the adjacent vessels. This proves that the water inside the plant represents a labrynthine system and any disturbances here and there are going to have only a local effect and not a general influence.

Another criticism against this theory which has not been satisfactorily answered, is that, the cohesive theory assumes tracheids to be more efficient than vessles in conduction. Dixon indeed beleived that the end walls of tracheids confer stability for the water system subjected to stress. If this were to be true, it would indeed be a paradox that the most dominant flora on Earth - the flowering plants have chosen a system of water conduction that is second best.

Barton Wright (1967) opines "It must be admitted that at the present time, there is no adequate explanation of the ascent of sap and although the cohesion of water possibly plays a considerable part in the matter, it is probably only a portion of the truth of the whole process".

Another point worth pondering is water absorption and upward movement are two processes that are of fundamental importance to the survival of the plant. Is it prudent for any organism to leave such processes to be controlled and opetated purely by physical forces,, not involving the biological system in any way? This is not to say that vital theories are correct, but only to point out that physical theories including cohesion theory are inadequate to explain the phenomenon of water uptake. While there is no doubt that physical forces do provide the mechanical energy for lifting the water up, there has to be some sort of a biological control. At least to that extent the participation of the living cells seem to be necessary. In what form and manner the living cells participate in water uptake; it is for further research work to find out. Any way for the present, the final word on the mechanism of ascent of sap has not been said.

7

TRANSPIRATION

Water is one of the most important compounds that not only helps as a medium for various biological reactions, but goes into the structural framework of the cell in plants. A large quantity of water is absorbed by plants from the soil in order to perform various biological activities, but one peculiarity is, most of this water is evaporated into the atmosphere and only a fraction is retained in the plant system. It is one of the oddities of nature as to why it makes a plant absorb large quantities of water and again forces it to lose much of it into atmosphere. This phenomenon of evaporation of water from the aerial parts of the plantbody is called transpiration. The fundamental question that has to be answered in water loss is, whether the plants absorb more water because they have to transpire or whether they transpire because they absorb more water.

When the atmosphere is not humid i.e., it is not fully saturated with water, there is a partial vacuum and water vapour escapes from the plant through the process of diffusion. Basically transpiration is a phenomenon of evaporation of water (from any exposed surface) but differs from it due to the involvement of the biological system. The following are the differences between evaporation and transpiration.

	TRANSPIRATION	EVAPORATION
1.	It is a physiological process	It is a physical process
2.	It is regulated by the opening and closing of guard cells	There is no biological control of evaporation from an exposed area
3.	Several types of biological factors are involved	No involvement of biological factors
4	Surface of the cells become wet due to transpiration (it prevents dryness of cell surface)	Surface becomes dry due to evaporation

Magnitude (quantity) of transpiration

In order to understand the physiological impact of transpiration, it becomes necessary to know as to what is the quantity of water lost from the plant. This is not uniform for all the plants and also for different parts of the plant. There are various physical and biological parametres governing the amount of water loss from the aerial parts of the plant. The rate of transpiration is generally high during summer. It has been estimated that in trees like *Betula* the amount of water lost due to transpiration during summer is as much as five times the fresh weight of its leaves. (This approximately amounts to about the loss of 15,000 litres of water per acre per day). The following table shows the amount of water loss in plants per unit area (100 sq cm at 20°C with humidity of 50%)

	Name of the plant	Loss of water per day (grams)
1.	*Hedra helix*	4.09
2.	*Begonia vulgaris*	4.57
3.	*Phaseolus vulgaris*	2.81
4.	*Coleus blumei*	5.06
5.	*Cucurbita pepo*	5.39
6.	*Vicia faba*	16.4
7.	*Helianthus annus*	12.0
8.	*Chrysanthemum frutescens*	16.35
9.	*Ficus elastica*	6.3

Some more calculations have been made about the amount of water lost by different plants by way of transpiration. The sunflower plant loses an average of about 56 kg of water in about 4 - 5 months time. The amount of water loss in *Crotalaria juncea* for the same period is 27 kgs. A birch tree with 2 lakhs of leaves loses on an average about 350-400 kgs of water on a summer day (The intensity of transpiration is usually expressed in grams of water lost per hour per square metre of leaf surface).

The above calculations which give a representative account of water loss will indicate the physiological importance of transpiration in the life of plants.

Transpiration coefficient : Transpiration is sometimes expressed in terms of fresh or even dry weight of the plant. The number of grams of water lost in accumulating one gram of dry matter is called *transpiration coefficient* or *transpiration ratio.* Transpiration ratio varies from plant to plant and also in the same plant under different conditions. It is 200 to 300 for *Sorghum* and other millets and 500 for alfa-alfa, while only 50 for pineapple. According to Slatyer (1967) the desert plant communities have the highest transpiration ratio, which may be as much as 5000.

Relative transpiration : Transpiration is similar to evaporation in that in both, water evaporates from an exposed surface. Both are controlled by a number of external factors, but transpiration is also controlled by a number of internal factors. This being so, when all the external factors are same, the rate of transpiration slightly varies from that of evaporation. This difference is called relative transpiration. It may be defined as the *ratio of the rate of transpiration per unit area from a plant to the rate of evaporation from unit area of a free water surface.*

Types of transpiration : All the aerial parts of the plant body without exception transpire; but their efficiency will vary. However the chief transpiring organ is leaf, even though stems, flowers etc., will also transpire. Hence the major type of transpiration in land plants is *Foliar transpiration.* In the leaf again, the epidemal cells, stomata etc., are the chief places where water loss occurs. Similarly in the stems transpiration takes place through lenticels. Based on these, transpiration is divided into three categories viz., *Cuticular transpiration, Lenticular transpiration* and *Stomatal transpiration.*

Cuticufar transpiration : The surface of leaves and very often stems also are covered by a waxy coating called cuticle. The cuticle is actually a layer that is supposed to prevent loss of water. That is why in plants exposed to sunlight, the deposition of cuticle is heavy, while in plants of the shaded area it is thin. However, water still evaporates through the cuticle and it is known as cuticular transpiration. Cuticular water loss however is not much when compared with the overall rate of transpiration. In barely it amounts to about 3-10% of total transpiration.

Lenticular transpiration : In woody plants, after the secondary growth, the epidermis is replaced by periderm whose outer layers are composed of dead cells (cork). In order to provide aeration to the inner living tissues, at certain places the cork tissue opens out and forms small slits called *lenticels.* Transpiration taking place through these lenticles is called *lenticular transpiration.* In comparison with other types of transpiration, this is very insignificant as it amounts to only 0.1% of the total.

Stomatal transpiration : Stomata are small openings present on lower, upper or both the surfaces of the leaf. By and large most of the water loss in plants (90-95%) is stomatal transpiration. Transpiration through stomata is known to occur in two stages. First, water evaporates from the moist cell walls of mesophyll cells into the intercellular spaces and secondly from the intercellular spaces into the exterior through the stomata. Various aspects of stomatal distribution, morphology etc will affect the rate of transpiration. We shall study these aspects in some detail.

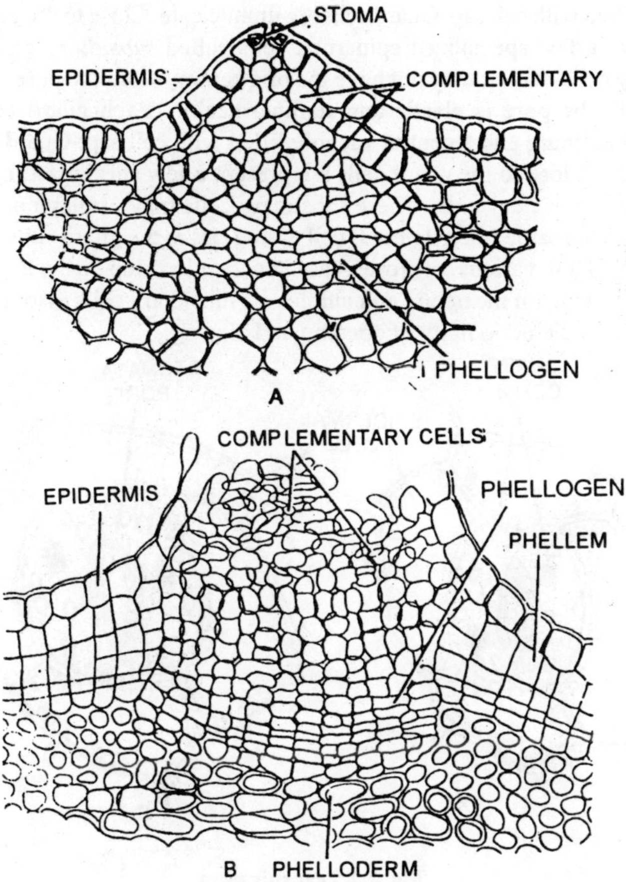

Fig. 7.1 Transpiration
Structure of a lenticel, **A.** Young, **B.** Mature

Structure of a stoma : The stoma is a minute pore or opening in the Epidermis of the leaf. The opening is usually surrounded by two elliptical (crescent shaped) cells called guard cells. These cells have a dense granular cytoplasm and some chloroplasts. The size of the stomatal aperture varies. It is 4 mm wide and 26 mm long in Corn. 3 X 7 mm in Phaseolus valgaris, 8 X 38 mm in *Avena* spp. etc. The cell walls of guard cells are unevenly thick and this plays an important role in the opening and closing of the stomata. Usually the wall facing the pore is thicker than the one away from the pore. The wall away from the pore is elastic and semipermeable. Each guard cell has a cytoplasmic lining and a central vacuole filled with cell sap. Guard cells are uninucleate. Close to the guard cells are present a few specialized epidermal cells called *subsidiary cells.* These subsidiary cells may or may not have the origin away from the pore. The wall away from the pore is elastic and semipermeable. Each guard cell has a cytoplasmic lining and a central vacuole filled with cell sap. Guard cells are uninucleate. Close to the guard cells are present a few specialized epidermal cells called *subsidiary cells.* These subsidiary cells may or may not have the origin from the same (meristemoid) cell giving riss to the guard cells. Stomata are classified into various categories based on their location on the leaf surface, on development, on the nature and number of subsidiary cells associated with it and also on the periodicity of opening and closing.

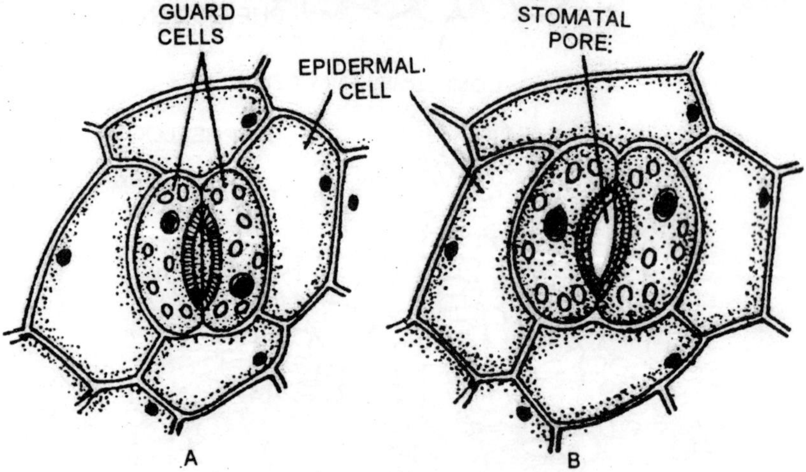

Fig. 7.2 Transpiration
Structure of a stomotal apparatus, **A.** Closed stoma, **B.** Opened Stoma

Classification of stomata

1. Nature and number of subsidiary cells : The following are the types of stomata.

(a) **Anomocytic or irregular celled type :** There are no subsidiary cells. It is also known as ranunculaceous type.

(b) **Anisocytic or unequal celled type :** Each stoma is surrounded by three subsidiary cells, of which one is smaller and the other two are larger.

(c) **Paracytic or parallel celled type :** Stoma is surrounded by two parallely arranged subsidiary cells. It is also called **rubiaceous type.**

(d) **Diacytic or cross celled type :** There are two subsidiary cells surrounding a stoma and they are at right angles to the guard cells.

(e) **Graminaceous type :** This is characteristic of the family Gramineae. The subsidiary cells are dumb-bell shaped.

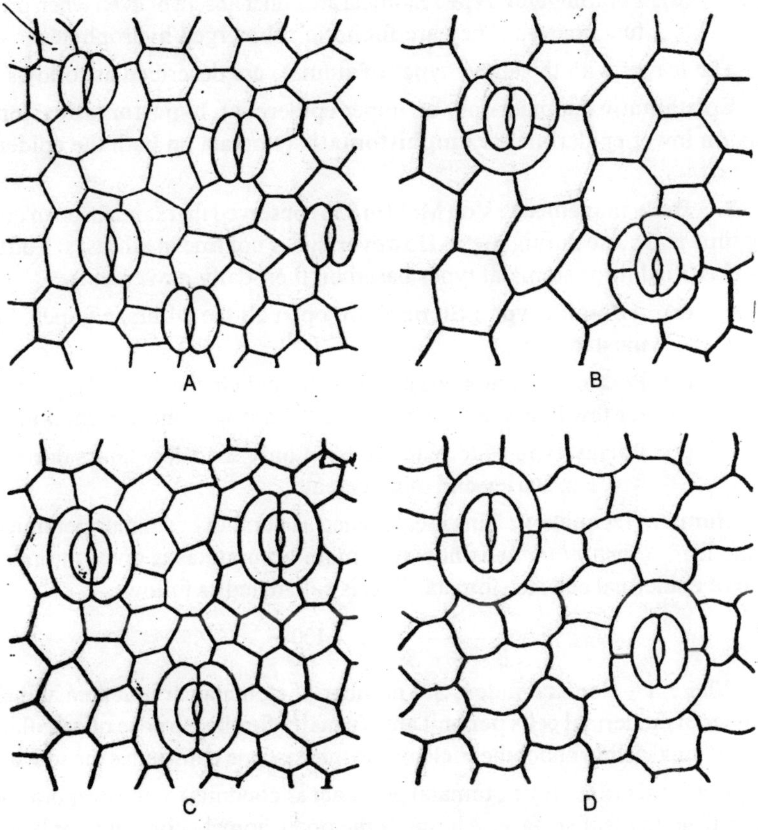

Fig. 7.3 Transpiration

Types of stomata, **A.** Anococytic, **B.** Anisocytic, **C.** Paracytic, **D.** Diacytic

2. Distribution : Based on their distribution the stomata are classified into -

(a) **Apple and mulberry type :** Stomata are found only on the lower surface of the leaf. Most common in trees.

(b) **Potato type :** Stomata are found on both surfaces, but they are more on the lower epidermis of the leaf than on the upper epidermis, e.g. Tomato, beans, pea etc.

(c) **Oat type :** Stomata occur on both surfaces of the leaf in equal proportions. They occur in isobilateral leaves, e.g. maize, corn etc.

(d) **Water lily type :** Stomata occur only on upper surface. This type of stoma is usually found in acquatic plants with floating leaves.

(e) **Potamogeton type :** Stomata are either absent or even when present functionless. They are found in submerged hydrophytes.

The leaves with the above types of stomata are described as follows -

Epistomatic (stomata only on upper epidermis), **hypostomatic** (stomata only on lower epidermis) or **amphistomatic** (stomata on both the epidermal layers).

3. Daily movement : Von Mohl (1856) observed that stomata open during day time and close during night. However this is not true in all cases. Lottfield has classified three stomatal types based on their daily movements.

(a) **Alfa-alfa type :** Stomata are open all the 24 hrs. e.g. pea, bean, mustard etc.

(b) **Potato type :** Stomata are open throughout the day and night except for few hours in the evening, e.g. cabbage, onion, pumpkin etc.

(c) **Barley type :** Stomata are open only for a few hours during the day. e.g. Barley and other cereals.

Stomatal frequency : This is calculated as number of stomata per unit area of the leaf. *Stomatal index* is however more accurate as it gives the relative ratio of epidermal cells to stomata. This is calculated as follows.

$$I = \frac{S}{E + S} \times 100$$

Where, I = stomatal index, S = number of stomata per unit area and E == number of epidermal cells per unit area (usually for the purpose of calculation, the leaf area visible under the Field of the microscope constitutes the unit area).

Stomatal efficiency : Stomatal pores act as conduits for the evaporation of water from the leaf surface. Although the open stomata constitute only about 1-2% of the total leaf area, the amount of water that evaporates often exceeds 50% of that evaporating from a free surface. Stafelt (1932) reports that a leaf of birch may have 60% or more transpiration in comparison with the free water surface of same measurement.

Brown and Escanbe (1900) described as to how diffusion through pores is much more efficient than diffusion from an open surface. According to them, diffusion through pore is nearly proportional to its perimetre than to its area. This has been subsequently supported by various workers such as Sayre (1926), 1\irrell (1936), Verduin 1949) etc.

The reason for a higher diffusion rate through pores than through a general surface is as follows - When water evaporates from a surface, the escaping vapours form a cloud or shell over the entire area. As a result, the vapour pressure gradient is reduced and only from the perimeter any appreciable diffusion can take place, because there will be reduced vapour pressure. In comparision with diffusion from the whole area, this perimetre diffusion is quite less. In the case of diffusion through stomata, over an area, hundreds of vapour shells are formed which do not overlap and thus maintain a lower vapour pressure. In this *case,* the perimeter diffusion will be much large in relation to the area as the perimeter of a small area is much larger than the area.

However in order to be very efficient, the stomata must not be too close, because the vapour shells might overlap. A normal vapour cap or shell is hemispherical and it can maintain this shape when it does not overlap on another vapour shell.

Stomatal movement : This is brought about by the changes in the shape and volume of the guard cells. It is beleived that turgor pressure plays an important role in this. The following experimental evidences suggest the involvement of a turgor mechanism.

(a) The isolated stomata open when floated on water, but close when floated on a hypertonic solution.

(b) The stomata on isolated epidermal strips behave exactly opposite than what is described in A.

(c) Behaviour of stomata in B resembles that of A when epidermal cells (subsidiary cells) next to guard cells are punctured.

Change in the turgor of guard cells is brought about by increase in the sugar content. As a result, water flows into the guard cells from the surrounding cells due to the increased OP (in guard cells). The increased water flow increases the TP of guard cells and the thin wall near the pore extends inwards resulting in the opening of the stoma.

Mechanism of transpiration : The major type of transpiration, which is of any consequences to plants is stomatal transpiration. Hence in the mechanism of transpiration, we will discuss aspects or forces that are responsible for the opening and closing of the stomata. Transpiration takes place when stomata open and it almost stops when stomata close. The following are the various views regarding the mechanism of stomatal movement.

1. Photosynthetic theory

Generally the stomata are open during day time and close during night. It was Von Mohl (1856), who thought that the photosynthetic activity in the guard cells is responsible for the opening of the stomata. According to him, the chloroplasts in the guard cells, photosynthesize during day time increasing their sugar content. As a result the OP of the guard cells increases and this attracts water from the surrounding cells. The guard cells increase their size due to inflow of water and the stomatal aperture opens. During night when there is no photosynthesis, there will be less sugar in the guard cells and the stomata close.

Objections

(a) In many instances, epidermal cells have more chloroplasts than the guard cells. This should result actually in the closure of stomata during day time as there will be increased OP in the epidermal cells due to

(b) Increasing the CO_2 concentration surrounding the leaf will increase the photosynthesis. But this seems to be having no effect on the opening of the stomata.

2. Starch - sugar hypothesis

Lloyd (1908) discovered that the amount of starch in guard cells increases during night and decreases during day time. The reverse is true for sugar content of the guard cells. According to him the starch sugar interconversion is responsible for the opening and closing of the stomata. He beleived that the change in the OP causes the interconversion of starch - sugar. Lottfield (1921) also discovered the same phenomenon in the guard cells, and he also reported that for the surrounding epidermal cells, the reverse (i.e. increased starch and decreased sugar content during day and increased sugar and decreased starch during night) is true.

Sayre (1926) observed that stomata remain Open when the medium is either alkaline or neutral even in dark and are closed in acidic pH even during day time. This means to say that the pH of the medium decides the interconversion of starch and sugar. According toSayre (1926) utilization of CO_2 during photosynthesis causes the increase in pH resulting in the conversion of starch to sugar. Scarth (1932), Small et al (1942) supported the idea of Sayre. Scarth (1932) proposed that during day time CO_2 liberated by respiration is utilized by mesophyll cells in photosynthesis, which results in lowering of H ion concentration. Whenever the pH is high the enzymatic conversion of starch to sugar is favoured. In darkness, CO_2 accumulates in the intercellular spaces and increases the H ion concentration and decreases the pH promoting conversion of sugar into starch.

The discovery of the enzyme *phosphorylase* in guard cells by Hanes (1940) greatly supported the starch sugar interconversion theory. Subsequently Yung and Tun (1948) Ymn and Willis (1954) have confirmed the participation of phosphorylase enzyme in the interconversion of starch and sugar in the guard cells. This can be represented as follows.

$$\text{Starch + Inorganic phosphate} \underset{\text{dark pH 5.0}}{\overset{\text{light + pH 7.0}}{\longrightarrow}} \text{Glucose 1. phosphate}$$

Starch is hydrolysed to form Glucose 1. phosphate at pH 7.0, but at pH 5.0 Glucose lTphosphate condenses into starch.

Steward (1964) proposed a scheme which is slightly modified. According to him Glucose 1. phosphate should have its phosphate removed to be ultimately able to increase the concentration of the cell sap. The process requires the participation of biological energy. The process of opening and closing of stomata according to Steward's scheme is as follows

Opening of stomata

$$\text{Starch + phosphate} \xrightarrow{\text{Phosphorylase}} \text{Glucose 1. phosphate}$$

$$\text{Glucose 1. phosphate} \xrightarrow{\text{Phosphoglucomutase}} \text{Glucose 6. phosphate}$$

$$\text{Glucose 6. phosphate} \xrightarrow{\text{Phosphatase}} \text{Glucose + phosphate}$$

Closing of stomata

$$\text{Glucose + ATP} \xrightarrow{\text{Hexokinase}} \text{Glucose 1. phosphate}$$

$$\text{Glucose 1. phosphate} \xrightarrow{\text{phosphorylase}} \text{Starch + phosphate}$$

Objections

Steward's theory has been criticised as follows.

1. Some guard cells (monocots) do not have starch, although they have other polysaccharides.
2. There is no evidence to show the presence of sugar at a time when starch disappears, when the stomata open.
3. In some instances starch is formed at a pH found during the opening of the stomata.
4. Often stomata close at mid day without any change in starch content.
5. The theory accounts for utilization of energy for closing, while opening does not require any energy.

Zeiitch (1963) proposed a theory that glycollic acid may be involved in stomatal opening. He showed that *a* hydroxysulphonates, which are competitive inhibitors of glycollic acid oxidase prevent the opening of the stomata during light. Addition of glycollic acid to the medium will reverse the process. Another evidence for Zeiitch's theory comes from the fact that low oxygen concentration inhibits both glycollic acid metabolism as well as stomatal opening. Similarly increased CO_2 concentration inhibits stomatal opening and glycollic acid formation.

3. Potassium (K^{++}) transport mechanism : Studies with the opening and closing of stomata, have shown that the active transport of potassium ions plays a critical role. Imamura (1943), Yamashita (1952), Fischer and Hsiao C1968) established a direct link between the stomatal movements and the transport of K^+ ions in the guard cells. According to Fujino (1959), the active movement of K^+ ions in and out of the guard cells causes the opening and closing of stomata. The work of Humble and Raschke (1971) on *Vlcia faba* fully support Fujino's theory. The theory can be briefly summarised as follows.

During light induced opening of stomata, K+ ions enter the guard cells from the surrounding epidermal cells, which act as storage of K ions. The transport of potassium is an active process involving the hydrogen potassium exchange pump at the surface of the plasma membrane. Many experiments have shown that in the ion exchange, the an ions of the organic acids (specially malic acid) provide H ions in exchange for K ions to balance the charges of K in the vacuole. Raschke 1975). The origin of malic acid at this stage can be explained by any one of the pathways.

a) During day starch is metabolised to malic acid.

b) Light induced release of malic acid from chloroplasts into the cytoplasm. (Levitt 1964).

The uptake of K^+ is an active process, the energy for which comes from respiration (A.P. Singh, and Srivastava 1973).

The closure of stomata is beleived to be caused by a passive movement of K^+ from the guard cells into the epidermal cells (or subsidiary cells). According to Zeigler and Raschke (1975), there is an active reabsorption of K ions into the epidermal cells.

The role of CO_2 concentration in K ion transport has also been investigated, it has been shown that, when CO_2 concentration in the vicinity of the leaf is lower than the natural atmosphere (during day), K^+ ions are transported into the guard cells, and when CO_2 concentration is high (during night) the reverse process operates. The apparent paradox in succulent plants in this context has been explaind by Nishida (1963). In these plants, stomata open only during night, when organic acids are stored (Normally this should have caused the closure of stomata). These plants carry out *crassulacean acid metabolism.* In succulent plants, due to incomplete oxidation of respiratory substrate CO_2 is not released during night. During day time, organic acids are decorboxylated, a large amount of 002 is released and stomata remain closed. The CO_2 released will help in photosynthesis even though the stomata are closed. This type of night opening of the stomata is called *Scoto active,* as against *Photo active* opening.

While it is true that CO_2 plays an important role in stomatal movement, there are still some aspects which need to be fully understood, some of these are:

(a) How CO_2 dissipates proton gradient across the plasma membrane of the guard cells.

(b) Guard cells have a kind of biological clock. In plants kept continuously in dark or light, stomata close and open according to the time of night or day. This biological rhythm dies out, of course in few days. What controls this rhythm?

(c) The observation of Dowes (1971), that the stomata on the upper and lower epidermis behave differently for the same amount of CO_2 intensity is yet to be explained.

(d) Humidity differences around the stomatal aperture affect its movement. For instance decrease in humidity encourages stomatal closure.

(e) When Abscisic acid is applied on to the surface of the leaves, stomata close. ABA appears to block the active excretion of H ~ from guard cells. Milborrow (1974) has proposed that abscisic acid prevents the uptake of K^+ or encourages the loss of K^+ from guard cells. But the mechanism is not fully understood.

Factors influencing the stomatal movement

The following factors are known to influence the opening and closing of stomata.

(a) CO_2 concentration : A high concentration of CO_2, in light or darkness brings about the closure of stomata. MansField (1965) has reported that stomata remain open even in darkness, if the CO_2 level in the surroundings is brought down.

(b) Light : Generally light has an effect on the stomata which tends to open the aperture wide. Blue and red rays of light seems to be most effective in the stomatal movement.

(c) Effect of water stress : Under conditions of stress, when mesophyll cells are starved of water, the result may be partial or complete closure of stomata. Water content is beleived to affect the stomatal movement in two ways - *hydropassive control* and *hydroactive control.* In hydropassive movement, the overall water potential (-8 to -11bars in *Phaseolus)* controls the stomatal movements. Hydroactive control is brought about by the synthesis of Abscisic acid (ABA). It has been observed that water stress leads to the formation of ABA, which inturn causes the closure of stomata.

(d) Temperature : At low temperature (O^0C) and at very high temperature (30C) stomata are closed. At moderate temperature suitable for other physiological activities stomata are opened.

Factors affecting transpiration

A number of external and internal factors affect the rate of transpiration. These are

(a) Light : This has an influence on the opening of the stomata. Stomata open wide with the increase in the intensity of light. However even diffuse light causes the opening of the stomata. In darkness stomata are closed. Light also increases the permeability of the cell membrane, making it easy for water molecules to move out of the cell.

(b) Wind : Velocity of wind is a very important factor in governing the rate of transpiration. In still air, the vicinity of the leaf gets saturated with water vapours and there is no scope for more water vapours to come out. On the other hand, if the wind is breezy, water vapours are carried away as soon as they come out creating a DPD in the vicinity of the leaf. Hence more vapours can diffuse out enhancing the rate of transpiration. The bending and twisting of the leaves and branches in high velocity wind forces the water vapours out of the intercellular spaces. But surprisingly it has been observed that at very high velocity, the rate of transpiration may be retarded. This is due to the cooling effect on the leaf surface, which increases the vapour pressure on the outside of the leaf. Another effect of wind at very high velocity is by extreme increase of water loss, it reduces the capacity of mesophyll cells to store water. When the cells become flaccid, stomata close.

(c) Temperature : An increase of temperature will increase the rate of transpiration, not by increasing the vapour pressure of atmosphere, but by increasing the vapour pressure of the intercellular spaces creating a pressure gradient. Generally increased temperature has an accelerating effect on all physiological activities. At very high temperatures however stomata close possibly because, the mesophyll cells become flaccid due to excessive evaporation. Initially the increase in the rate of transpiration follows Vant Haff's rule up to certain extent ie for every 10^0 rise in temperature, the rate of transpiration gets doubled.

(d) Atmospheric pressure : Generally a reduction in atmospheric pressure, reduces the density of water vapours and thus allows rapid diffusion of water vapours. Plants at hill tops show higher rates of transpiration due to this and also due to the velocity of wind. But this increase in the rate is to some extent offset by low temperature.

(e) Atmospheric humidity : The amount of atmospheric humidity directly affects the rate of transpiration. The relative humidity of the atmosphere is a vital factor in humidity influence on transpiration. (Relative humidity is the difference between the actual humidity and the saturation humidity in that area). The quantity of relative humidity and rate of transpiration are inversely proportional. The difference in vapour pressure inside the leaf and outside the leaf is the one that actually governs the rate of transpiration.

(f) Availability of water : Adequate amount of water in the soil in which the plant grows is a pre requisite for transpiration. Reduction in soil water, reduces its absorption and hence reduces transpiration. In saline soils also, transpiration is less, possibly because salinity affects absorption of water.

(g) Structural attributes of plant : Stomatal frequency, size, position and posture of the aerial organs, deposition of cuticle, development of hairs and a host of other morphological and anatomical features affect the rate of transpiration. Usually most of the structural peculiarities are to be found in xerophytic plants. Leaf fall during summer noticed in deciduous plants is-also an adaptation to check transpiration.

Significance of transpiration

Whether transpiration is advantageous or disadvantageous, whether it is necessary or not is being debated at length. There is no unanimity among plant physiologists as to the benefits or harm caused by transpiration. We shall discuss this in some detail. Before discussing as to whether transpiration is necessary or not for the plants, we shall try to list its so called advantages and disadvantage.

Advantages of transpiration

1. It creates a suction force to absorb water from the soil.
2. The main force in ascent of sap is brought about by transpiration pull.
3. It affects DPD and thus helps diffusion through the cells.
4. Transpiration helps in mineral uptake by developing suction force.
5. Excess quantity of water is generally absorbed by the plant in order to take in the required quantity of mineral salts. If this exess of water is retained in the plant, it will not only disturb the osmotic balance between cells, but may also bring about decay of the tissues.
6. Transpiration stream helps in the translocation of solutes from one part of the plant to another part.
7. Metabolic activities in the plant increase the internal temperature (of the plant). Transpiration brings down the temperature by allowing fresh water to flow through the length and breadth of the plant.

8. The opening and closing of stomata during transpiration indirectly influence respiration and photosynthesis.
9. It enhances the rate of movement of molecules, so that they can easily and speedily migrate from one part to another part. For instance a molecule of magnesium sulphate, perhaps would require a hundred years or more if it were to move by its own kinetic energy and reach the top of a 300 feet high tree. But by the mass movement of water brought about by transpiration it can move at a much faster rate.

Disadvantages of transpiration

It is widely beleived that transpiration helps to remove excess of water. This is not true because, in the absence of transpiration perhaps, the plants would not have absorbed excess quantity of water. Now plants absorb water far in excess of their requirement, because they have to lose bulk of it (water) by way of transpiration. In some soils with inadequate water, excess transpiration often kills the plant, by i removing all the water. It has been found out that plant cells can maintain their turgidity even in the absence of transpiration.

Very often it is argued that it is transpiration that helps in the increased. absorption of water and together with it mineral salts also are absorbed. This is not true, for mineral salt absorption is an active process independent of transpiration. There is no mass flow of salts into the root hairs along with water. It is true however that mineral salts are translocated in xylem under the influence of transpiration stream. But as recent evidences have indicated, a major part of the salts are translocated through phloem. If transpiration were to be responsible for translocation of salts, there should be more amount of salts in the ash of plant tops than roots. But it is not the case.

The role of transpiration in reducing the internal temperature of plants is a dubious one, as leaves in which the transpiration is artificially checked (by the application of vaseline) do not show any undue increase of temperature. Desert plants which have adaptations to check transpiration do not show any harmful effects of increased temperature. The actual heat dissipation is brought about by radiation of extra heat into the atmosphere. The various types of adaptations to check transpiration seen particularly in desert plants is a clear indication of the harmful effects of transpiration. In a desert environment, where the outside temperature is very high, the internal temperature of plants is also likely to be more and if transpiration were to play any role in reducing internal temperature, plants would have actually promoted transpiration, rather than checking it.

After all these, a question naturally arises as to why plants have retained transpiration inspite of all its ill effects on the biological system as it poses an unnecessary burden of absorbing large quantities of water. Plants should have dispensed with transpiration. If there were to be a choice perhaps, plants would have gladly done so, but the route of transpiration also happens to be the pathway for many other vital processes required for the survival of the plant (e.g. photosynthesis, respiration etc). The presence of stomata, loosely arranged mesophyll cells, are all necessary for gaseous exchange during respiration and photosynthesis. Transpiration is unavoidable because of this peculiarity of anatomical construction of the leaf which facilitates evaporation of water in a system originally meant for gaseous exchange. Transpiration can thus be said to be an implication or unforeseen consequence of leaf anatomy. As long as ingress, and egress of gases continue, transpiration will also continue. From this point of view, in order to retain a suitable anatomical framework of tissues in leaf, transpiration is necessary even though it is an evil by itself. *It is thus a necessary evil.* (Curtis, 1926)

Antitranspirants

As the very name indicates, these represent artificial devices introduced by man to check transpiration. This is true because transpiration does not have salutary effects on the plant. But checking transpiration is easier said than done, because any physical or physiological check on transpiration will directly affect gaseous exchange, thus disturbing photosynthesis and respiration. If this were not to be the case, perhaps checking transpiration would have been very easy. A mere foliar application of vaseline would check stomatal transpiration.

If indeed an antitransparent device could be found out, it effectively reduces water requirement by plants to more than half of the present rate. An ideal antitranspirant would be one which checks transpiration and at the same time does not disturb the process of gaseous exchange.

Among the antitranspirants thc ones that are in use are colourless plastics, silicone oils and low viscocity waxes that are sprayed on leaves to form a Filmy barrier which can be negotiated by gases but not by water vapours. So far only limited success, has been achieved using these antitranspirants. A second group of antitranspirants are chemicals which when sprayed on leaf)cause partial closure of stomata. These substances however should be effective only against guard cells, but should have no influence on epidermal cells. To date only one chemical qualifies as an antitranspirant in this category. The fungicide *phenyl mercuric acetate,* which when sprayed in a concentration of 10 M, is almost non toxic to the leaves and causes partial closure of stomata for up to two weeks. Spraying of Abscisic acid, a natural growth regulator is also known to cause stomatal closure. But the effect wears out too soon to be of any practical use.

CO_2 is another effective antitranspirant. Raising the level of CO_2 from its normal 0.03% to 0.05% is known to cause closure of stomata. High concentration of CO_2 should be avoided for not only does it close stomata, but will also have deleterious effect on phtosynthesis. Increase of CO_2 content in possible however only in green houses.

To summarise, it is true that at present large scale application of antitranspirants is not feasible both because of their impracticability and absence of total success. It can be expected that continued reserach in this exciting field of applied plant physiology will create a revolution in agriculture by minimising the dependence of crop plants on water.

Guttation

Under certain environmental conditions, water comes out through the aerial parts (leaves) in the form of drops (not in vapour form). This physical loss of water is called *guttation.* Wheh soil conditions favour, a high rate of water absorption and the rate of transpiration is low, guttation takes place. The term guttation was First used by Burgerstein, to describe loss of water from the leaves in liquid form.

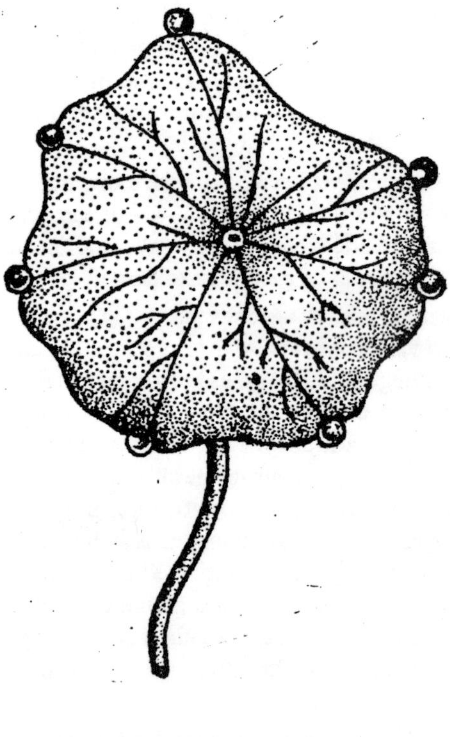

Fig. 7.4 Transpiration
Leaf of Nasturtium showing guttation

The magnitude of guttation in comparison with transpiration is very low, . besides guttation is not universally observed. It has been reported in 333 genera belonging to 115 families. Even in these plants guttation occurs depending on the environmental conditions. *Colocasia antiquorum* is known to guttate upto 100m' of water per day. Guttatian is observed, early in the morning, after a warm and humid night. Drops of water adhering to the tips or edges of leaves are often seen in grasses, tomato, *Nasturtium* etc, which are often mistaken for dew drops.

The exudation of water in guttation is due to positive root pressure. Just as transpiration takes place through stomata, guttation occurs through special structures called *hydathodes,* which are often called *water stomata.* Hydathodes are seen at the tips and margins of leaves. This is the reason why, water drops are seen at the tips and margins of leaves.

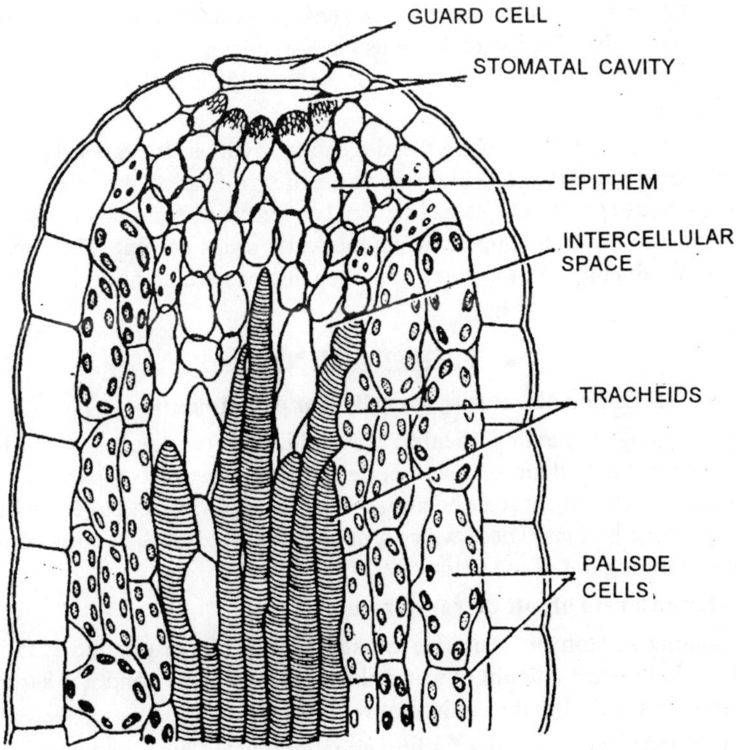

Fig. 7.5 Transpiration
Structure of a Hydathode

Structure of a hydathode : Each hydathode opens to the exterior by means of a pore in the leaf epidermis. Internal to the pore is the epithem, which..is made up of loosely arranged thin walled parenchyma cells. The epithem usually occupies the tip of a vein. Surrounding the epithem is a mass of chlorenchymatous tissue.

Haberlandt, has recognised two types of hydathods viz., **epithem hydathode** and **active hydathode. In** tne rormer water loss is by the force of root pressure. While in the latter, water is removed by the force developed by the cells of the epithem.

As has already been pointed out, root pressure may be the main force responsible for guttation. Conditions that discourage root pressure such as cold, dry aerated soil etc., reduce guttation also. Mineral deficiency also is known to affect guttation.

Guttation water is not pure in the sense, it is known to contain various compounds in dissolved state. Various kinds of enzymes, sugars, arnino acids, vitamins and mineral elements such as P, K, Na, Mg, Fe etc have been found in guttation water.

Guttation sometimes shows endogenous rhythms. It ceases during day and nearly always occurs at night. In terms of significance, guttatian is of no consequence to plants. Occasionally, it may even cause injury to the leaves, by salt deposition, which happens when the water drops dry out. This creates a favourable situation for the entry of pathogenic bacteria, fungi etc.

EXPERIMENTS

1. Demonstration of transpiration (Bell jar experiment)

Cover a pot containing a healthy plant with a bell jar. Apply vaseline to the lower rim of the bell jar to ensure perfect sealing. Keep the apparatus in sun light. After sometime it may be noticed that water drops get collected on the interior of the bell jar. These water drops are formed by the cooling of water vapours coming out of the plant.

2. Measurement of rate of transpiration

Ganong's potometer is used to measure the rate of transpiration by indirect method. In this, the amount of water absorbed is measured and it is assumed, this amount is equal to the amount of water transpired.

The apparatus consists of a horizontal graduated tube, which is bent in opposite directions at the ends. One bent end is wide and the other is narrow. A reservoir is fixed to the horizontal tube near the wider end. The reservoir has a stop cock to regulate the water flow. The apparatus is filled with water using the reservoir. A twig or a small plant is fixed to the wider arm through a split cork. The other bent end of the horizontal tube in dipped into a beaker containing coloured water. An air bubble *is* introduced into the graduated tube at the narrow end (arm). As transpiration takes place, the air bubble will move from one end of the graduated tube to the other. After it has reached near the reservoir (reservoir) is opened to push the air bubble back and then its (air bubble) movement can be watched again.

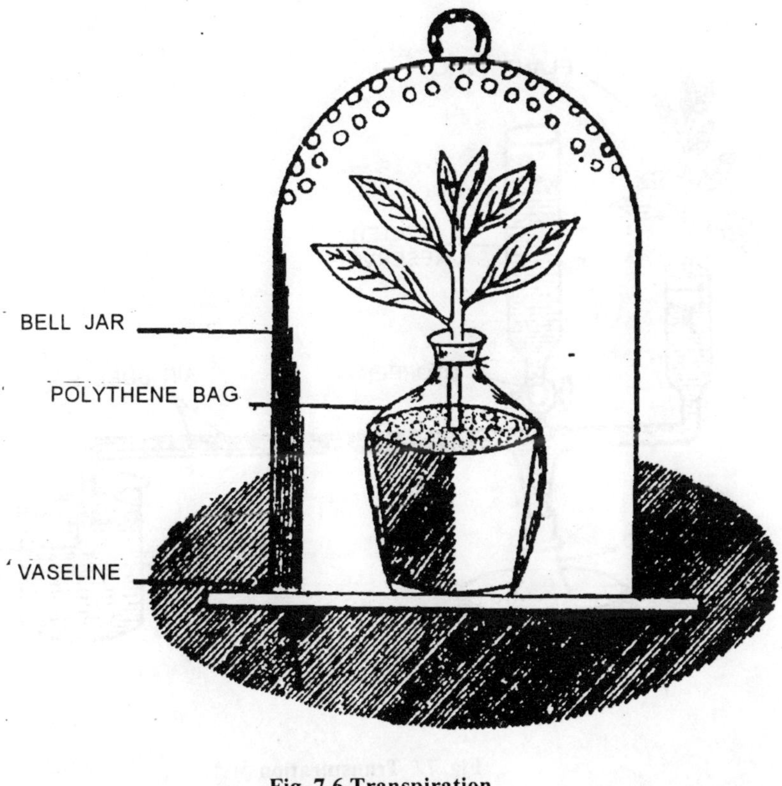

Fig. 7.6 Transpiration
Bell jar experiment

A record is made of the time taken to travel a definite distance in the graduated tube. Many observations are made and averaged to obtain the relation between distance and time. By using a simple formula, the amount of water transpired may be calculated $V = r^2 1$ (V = volume of water, = 22/7, r = radius of graduated tube, 1 = air bubble).

3. Four leaves experiment

(To compare cuticular and stomatal transpiration) Four leaves are hung on to a thread connected to two stands. Of these four, one is applied on the upper surface, the third has vaseline on only the lower surface and the fourth has vaseline on both surfaces. The apparatus is kept in sun light for few hours and observations are made. The leaf having vaseline on both surfaces does not dry, the leaf having vaseline only on upper surface dries faster than the leaf having vaseline on the lower surface. The leaf without vaseline dries faster than the leaf having vaseline only on the upper surface.

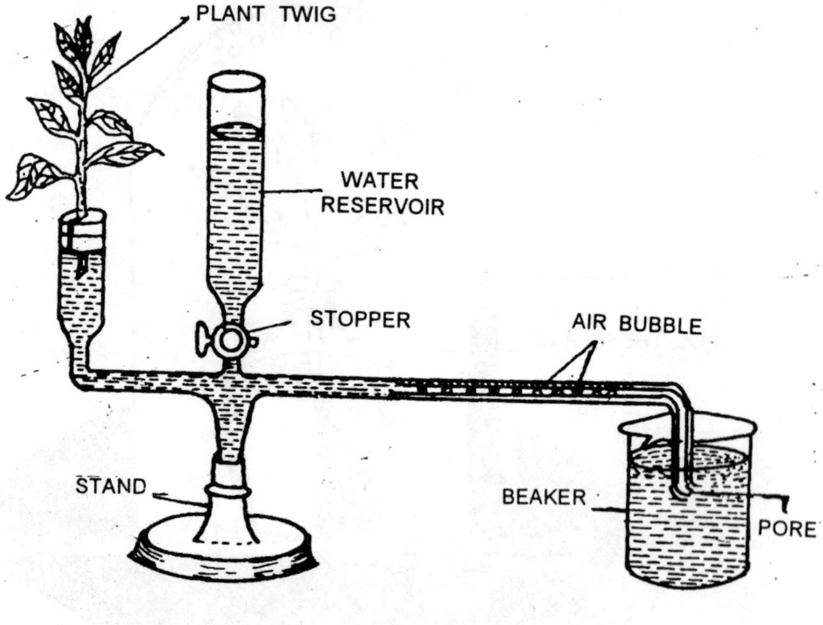

Fig. 7.7 Transpiration
Ganongs photometer

Since the leaves (in this case) have stomata only on the lower surface, the leaf having vaseline on lower surface does not dry soon, so also the leaf having vaseline on both surfaces, while the others dry up. This is an indication of stomatal transpiration being more than cuticular transniration.

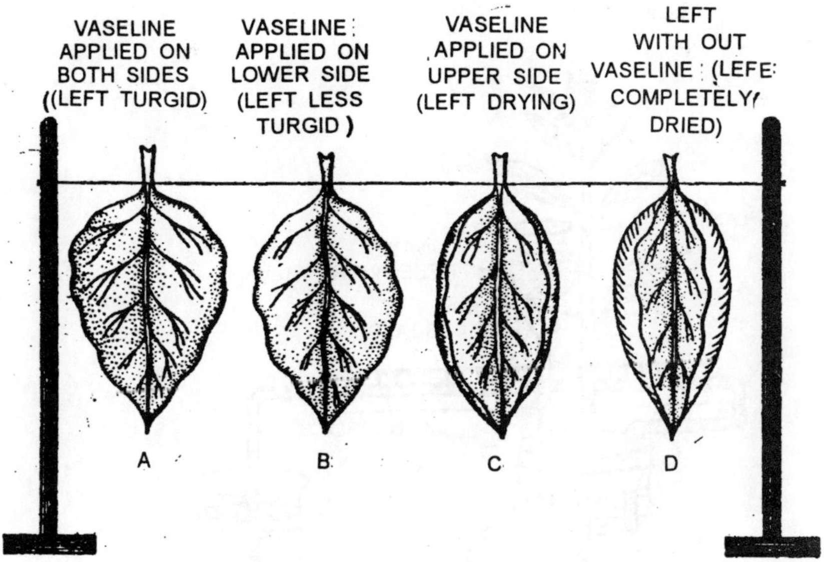

Fig. 7.8 Transpiration
Four leaves experiment

4. **Measurement of transpiration by Farmer's potometer**

The Farmer's potometer consists of a wide mouthed bottle. The mouth is fitted with a cork having three holes. One hole has a thrice bent tube, with a scale fixed to its horizontal length. The second hole has a reservoir with a stopper and through the third hole is fixed a twig.

Fill the apparatus with water and insert a healthy twig cut under water. Introduce an air bubble into one of the arms of the horizontal tube and immerse it in beaker containing coloured water. Apply vaseline to the edges of the cork to make it airtight. Keep the apparatus in sun light.

As transpiration goes on, the air bubble will move from one end of the tube to the other. The rate of transpiration can be measured by calculating the time against distance travelled by the air bubble.

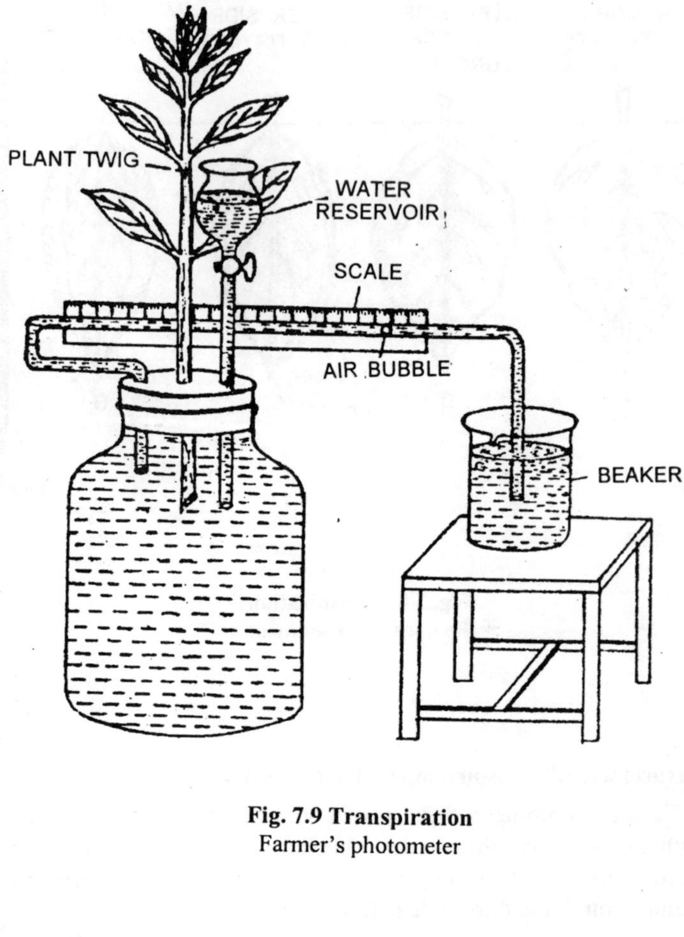

Fig. 7.9 Transpiration
Farmer's photometer

8

MINERAL SALT ABSORPTION

Plants depend on soil for the supply of raw materials necessary for their survival. Even though plants can manufacture organic food by themselves, they require water and inorganic nutrients as raw material. Soil has large amount of mineral salts dissolved in soil water. This constitutes the soil solution. Some solution. Some mineral salts in their ionic form are adsorbed *to* the soil micelles; *these* are also available for the plants.

Soil solution : The main constituents of soil are the clay particles or soii micelies. and these arc present in colloidal form. The micelles are negatively charged and they hold the positively charged ions on their surface in order to balance the electrical charge.

Soils are basicaily of two types, acidic and alkaline. In the former the clay particles are mainly associated with H+ ions and in the latter Ca^{++} is the principle cation associated with the claymicelis. Besides the above, clay particles also bind Potassium, Magnesium, Ammonium and many other cations. Many anions are also found in the soil. These are - $C1^{-}$, $So4^{-}$, HCo^{-}_{3}, $H2Po4^{-}$, $No3^{-}$ and oH^{-}. The ions that are loosely adsorbed to the soil micelles get easily exchanged to other ions of equal charge. Some ions have greater affinity (adsorption capacity)' to the soil micelles and these, cannot be easily exchanged. Generally, the following is the cation retentive capacity of the soil miceils.

$$H^{+} > Ca^{++} > M^{g++} > K^{+} > NH4^{+} > Na^{+}$$

It has been shown that both electrolytic and non electrolytic elements enter into the plant system through the root hairs. While non electrolytes enter as molecules electrolytes enter as ions. Entry of one type of ion into the cell has to be balanced bv the entry of another ion of equal electrostatic charge. Monovalent cations like K^{+} can enter at a much rapid rate than di or polyvalent cations (Ca^{++}, Mg^{++} etc) A similar preference is seen towards anion accumulation also in plants.

The cell interior (cell sap) where ions get accumulated is electrically neutral. In some instances, like the preferential accumulation of either cations or anions, the electrical stability gets disturbed. In such situations, ionic changes take place in the cell to balance the charge. If cation uptake is more than anion, the cell produces organic acids, the anions of which remain in cell sap to balance the excess cations. The H^+ ions of organic acids pass out into the external medium to compensate for the removal of cations.

The shortage of cations or anions in the cell interior (which may occur frequently) are balanced by the easily available H^+ and OH^- ions of water.

Site of mineral absorption in roots : Mineral salts are absorbed at the terminal regions of the roots which are in contact with soil solution. Unlike absorption of water, which takes place by root hairs, mineral salts are absorbed by the cells of epiblema. It is not yet definite as to which zone at the root tip is primarily involved in mineral salt absorption. The root tip consists of, root cap, meristematic zone, elongation zone, root hair zone and cell maturation zone. Some physiologists are of the opinion, that the zone of elongation and the zone of maturation are the principal sites of mineral salt absorption (Bar Yosef et al 1972). On the other hand Ney and Tinker (1977) opine that, the entire surface of the root tip is involved in mineral salt absorption. Some of the experimental findings indicate that while the entire surface of root tip might be involved in salt uptake, it is not uniform for all the salts. For instance in barley roots, phosphate, potassium and ammonium ions are absorbed equally from all areas; absorption of calcium is resticted to the apex, while iron is absorbed at the region of elongation.

MECHANISM OF SALT ABSORPTION

Early workers in plant physiology assumed that entry of mineral salts posed no problem as they enter into the root system along with water, when absorption of water takes place. They also assumed that mineral salts are translocated together with the ascent of sap in the xylem elements. The assumption of mass entry of salts into the roots, or their entry based on a positive gradient have been seriously questioned in recent years. Even as early as 1900, Pfeffer questioned the assumption of passive entry of salts along with water and observed "the nature of plasma is such as to render it possible that a substance may combine chemically with the plasmatic elements, thus being transmitted internally, and then set free again".

The various view points on the mechanism can be grouped under two headings -*passive uptake* and *active uptake*. The passive uptake theories beleive that salt absorption is purely a physical process controlled by concentration gradient, mass flow etc., and it is completely independent of respiratory energy. On the other hand, the active uptake theories hold that, the salts are accumulated preferentially against a concentration gradient indicating clearly the involvement of metabolic energy.

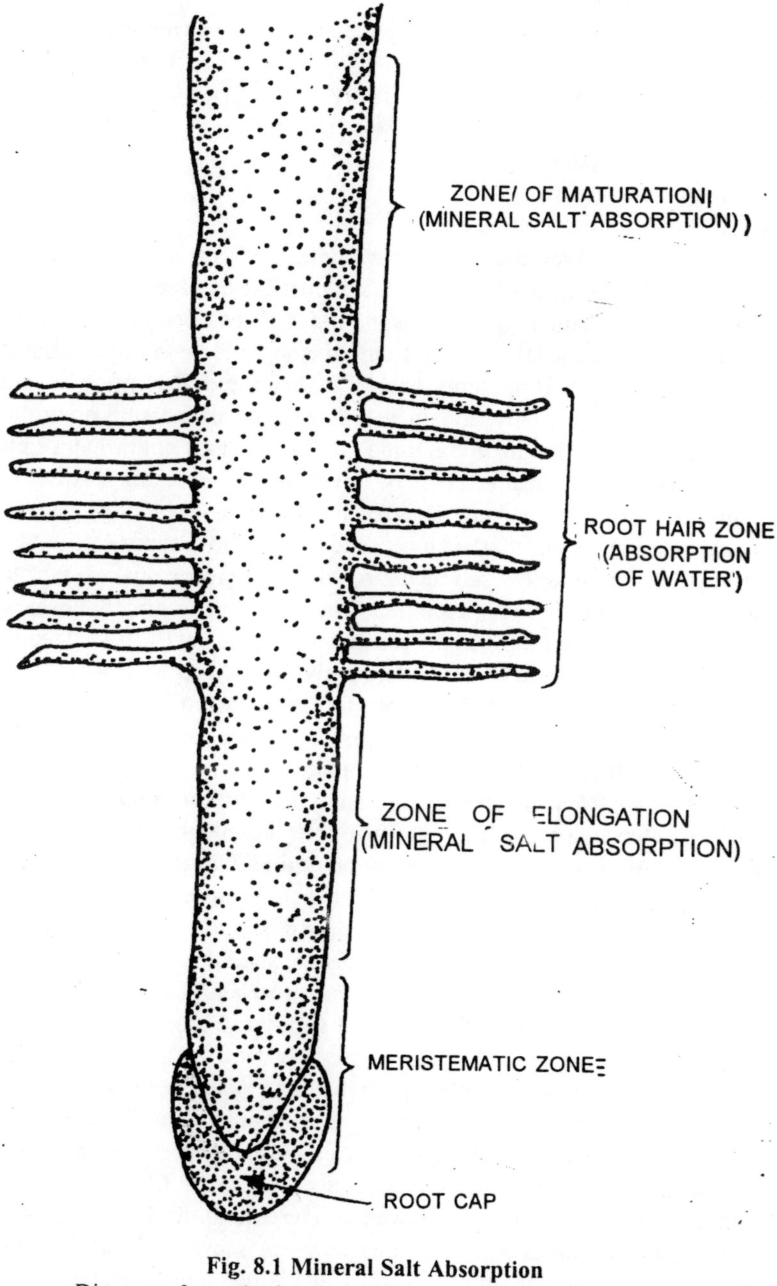

Fig. 8.1 Mineral Salt Absorption
Diagram of root tip showing the site of mineral salt absorption

PASSIVE ABSORPTION

Various physical forces are contemplated as playing a role in the passive absorption of salts. These are-

(a) Outer and apparent free space theory
(b) Mass flow theory
(c) Diffusion theory
(d) Ion exchange theory and
(e) Donnan equilibrium theory

Apparent free space theory : Passive uptake of salts under conditions, when participation of metabolic energy is not possible has been demonstrated by many workers (Briggs and Robertson, 1957). It has been observed that when a plant cell or tissue is transfered from a medium of low salt concentration to a highly concentrated medium, there is a very rapid absorption of salts initially which tapers gradually untill there is a steady uptake possibly controlled by metabolic forces. That the initial rapid intake of salts is not influenced by metabolic energy can be ensured either by low temperature treatment or adding metabolic inhibitors to the medium. Thus the initial rapid uptake is non metabolic. It has also been observed that such a tissue which had a rapid salt uptake, when transferred on to a low salt medium, will readily give out many of the accumulated ions. This means, then, the salts that were originally absorbed rapidly were in a free condition in the cell and they came out (diffused) readily into the external medium. In other words, a part of the cell or tissue where these ions were held and were given out is open to free diffusion. This is called the *outer space* or *apparent free space.* Apparent free space has been described as "an appreciable fraction of the volume of plant tissue or cells which can be penetrated by ions from an external solution by free diffusion and exchange". The volume or the area of the free outer space may be calculated by placing the tissue in a solution of known concentration and allowing salt uptake up to equilibrium. If we assums at this stage, that the concentration in the outer and inner medium or same, and knowing the amount of salt absorbed, we can easily calculate the volume of the free outer space. Hope and Stevens (1952) working with bean root tips found that an equilibrium is reached in about 20 minutes, when they (root tips) were placed in a KCI solution. Further work by Hope (1953) showed that the volume of the free space is not fixed and it may change according to the concentration of the outer medium. When the root tips rapidly absorbed salts from KCI solution initially, there should be no further passive absorption in the absence of involvment of metabolic energy. Hope (1953) showed that, at this stage, increase of concentration of KCI solution and addition of metabolic inhibitors, showed an increase in the free diffusion of ions. From this it becomes clear, that the free space is apparent and flexible in volume, so that it can accommodate salts inside the cell depending on the external concentration. Hence the term *apparent space* replaced the outer space (Hope, 1953).

The question now to be answered is what are the physical forces involved, that drive the salts inside the cell, in the absence of metabolic energy. Diffusion, mass flow, ion exchange and Donnan equilibrium are some of the mechanisms believed to be involved in the passive uptake of salts.

Mass flow : It has been observed that rapidly transpiring plants contain a high quantity of mineral elements. Further, increased rate of transpiration is known to accelerate the absorption of salts. When transpiration is taking place at its maximum pace, a large quantity of water flows into the root drawn in by the suction force. During this mass flow of water, large quantities of salts also get into the root system. Hylmyo (1953), Kramer (1956), Russel and Bather (1960) beleived in the mass flow of ions along with the transpiration stream. Lopushinsky (1964) showed a mass entry of salts in detopped tomato plants, when increased transpiration caused increased salt uptake. Many others believe that the effect of transpiration on salt uptake is indirect in that, by removing the salts from the xylem of the root, it creates a positive concentration gradient for the entry of salts.

But mass flow hypothesis of salt entry does not account for preferential accumulation of certain types of ions in the cell interior. It is quite possible that some amount of dissolved salts may enter the root along with water when conditions are favourable. But the whole of salt absorption cannot be explained by mass flow.

Diffusion : When the outer solution is highly concentrated with reference to salts than the interior, salts migrate into the cell on the principle of diffusion. This is true even when the cell interior is highly concentrated with reference to some but less concentrated with reference to some others. In such a situation different salts enter into the cell based on their partial pressure according to the principle of independent diffusion. When there is a favourable concentration gradient, salts diffuse into the cell even when metabolic inhibitors are introduced in the external medium.

Ion exchange : In this process, cations or anions inside the cell are exchanged with ions of equal charge from the external medium. Since this involves exchange of ions of similar charge there is no net gain or loss either to the plant or to the soil.

Ion exchange has been experimentally confirmed with barley roots. Excised barley roots kept in a solution of radioactive KBr, absorb radio active K^+ ions. When these roots are transfered to water, little or no loss of radio active ions into water is noticed. The membrane behaves as if it is impermeable to K^+ ions. When these roots are transfered to a non radio active KBr solution, the radio active K^+ ions move out into the external solution and simultaneously, the non radio active K^+ ions move into the root cells. The cytoplasmic membmae now allows movement of K^+ ions in either direction. Since the total amount of K^+ ions in the external medium remains unchanged, it is obvious that the radio active and non radio active K^+ ions have exchanged places. This ionic exchange can take place even in anaerobic conditions indicating the non involvement of respiratory energy. It appears a similar mechanism operates between the root exterior on the one hand and soil solution and the clay micelles on the other. The cations and anions of the soil micells (K^+, Mg^{++}, Ca^{++} etc, and Cl^-, Br^- etc.) may get exchanged with H^+ and $OH^=$ ions that are adsorbed on the root surface. H^+ and OH^- ions are readily available from water. While there is no net gain of ions, ion exchange provides a non active mechanism for plants to accumulate ions of the desired type. There are two means by which ion e~hange can take place. These are - **Contact exchange** and **Carbonic acid exchange**.

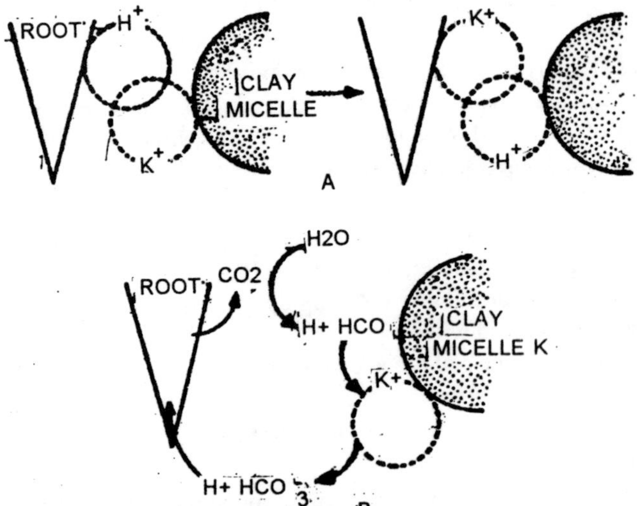

Fig. 8.2 Mineral Salt Absorption

Diagrammatic representation of **A.** Contact exchange, **B.** Carbonic acid exchange

Contact exchange : A direct exchage of ions takes place betweefa the root surface and soil micell. The ions that are adsorbed either to the root surface or soil micelle are not fixed, but oscillate, in a small volume of space. This is called the *oscillation volume*. When the root surface and the soil surface are very close, the oscillation volumes of the two ions adsorbed to them overlap and they exchange positions. For instance the root surface has H^+ and the soil micelle has K^+. On overlapping, H^- is transfered from root to soil micelle and K^+ from soil micelle to root. In this way, the root can exchange H^+ or OH^- ions to the required ions. This is called contact exchange, since the soil micelle and root come into direct contact and exchange the ions.

Carbonic acid exchange : In this mechanism, the soil micelle and the root exchange their ions *via* the soil solution which acts as a mediator. Roots release into the soil water CO_2 as a by product of respiration. This CO_2 combines with the soil water and forms carbonic acid (H_2CO_3). In the soil solution, the carbonic acid dissociates into hydrogen (H^+) and bicarbonate (HCO^3) ions. The carbonic acid exchange takes place at two stages. At the first stage the H^+ ions of soil solution will exchange their position with the K^+ ions of the clay micelle. In the second stage, the K^+ of the soil solution is transfered on to the root surface in exchange for H^+.

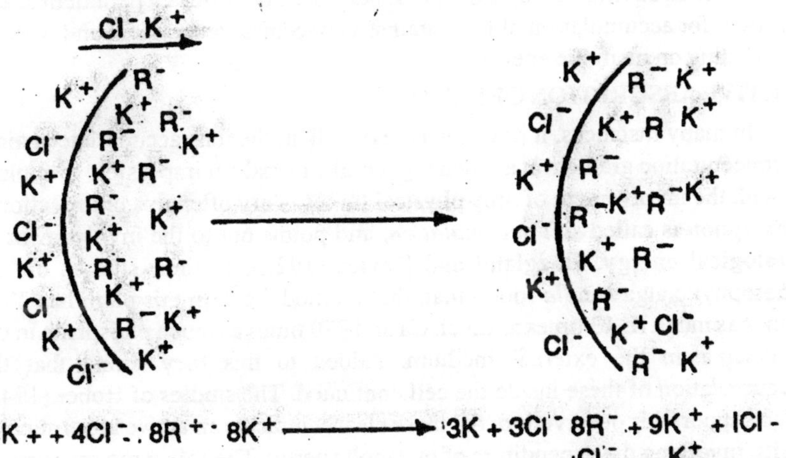

Fig. 8.3 Mineral Salt Absorption
Diagram to demonstrate Donnan Equilibrium

Donnan equilibrium : This theory is based on the assumption that certain ions, after they get into the cell interior become Fixed to the inner side of the membrane and thus are not freely diffusible. Donnan equilibrium is maintained by the co-operation of electrical and diffusion gradient to balance the electro chemical equilibrium. Let us take an example. A cell having potassium is immersed in a solution of potassium chloride. The cell interior has K^+ and Pr^-

ions and the external medium has K^+ and Cl^- ions. Of these Pr^- ions become fixed inside the cell and plasma membrane behaves impermeable to these ions. But K^+ and Cl^- are free ions and they diffuse easily. In such a system Cl^- moves into the cell due to its own concentration gradient (there is no Cl^- inside the cell) together with Cl^-, K^+ ions also enter the cell to maintain electrical neutrality. This will continue till equilibrium is reached.

At the equilibrium, the concentration of K^+ inside the cell is more than that of the external medium, because in order to neutralize the effect of non diffusible Pr^- more number of K^+ ions are required. As a result, K^+ ions can migrate into the cell against a concentration gradient without the expenditure of metabolic energy. This type of migration can take place for anions also, if there are fixed cations inside the cell. This kind of movement of anions or cations (from cell exterior) against their own concentration gradient in order to neutralize the effect of fixed cations or anions (inside ihe cell) is called Donnan equilibrium. Several such equilibria may exist in the cell based on different parametres, and account for accumulation of ions against a concentration gradient and without depending on metbolic energy.

ACTIVE ABSORPTION OF SALTS

In many instances, it has been observed lhat, the ions accumulate against a concentration gradient at a much higher rate, to make it impossible to explain it with the involvement of only physical forces. Very often this unidirectional absorption is called *salt accumulation,* and points out to the involvement of biological energy. Hoagland and Davies (1923), in their studies on the charophytic alga *Nitella* found that, the internodal cell in this plant has 2000 times as much K, 13 times as much Ca and 870 times as much phosphate in the cell sap as in the external medium. Added to this they found that the accumulation of these inside the cell continued. The studies of Hober (1943) also support this observation. All these clearly indicate an active absorption of salts, involving the expenditure of metabolic nergv. The following are some of the evidences for the existence of active absorption and its link with respiration.

1. The rate of absorption is too rapid to be accounted by passive absorption.
2. Mineral salts are accumulated against a concentration gradient.
3. The rate and quantity of absorption is directly related to metabolic activities.
4. There is a direct relation between salt absorption and respiration.
5. Respiratory inhibitors inhibit salt uptake also.
6. Salt induced respiration is the result of increased salt uptake.
7. Chemicals such as, dinitrophenyl hydrazone, dichlorophenyl dimethyl urea which inhibit ATP synthesis also inhibit salt uptake.

8. Actively dividing cells with a high rate of respiration also have a high rate of salt uptake.
9. Young roots immersed in salt solution, through which nitrogen is bubbled fail to absorb salts, and they renew salt uptake when oxygen is introduced into the medium.
10. Absorption of salts is indirectly linked to photosynthesis as it (photosynthesis) provides the respiratory substrate. The following studies clearly link respiration with salt absorption.
    (i) Steward (1932) and Hopkins(1956) observed a direct relationship between oxygen content in the medium and the rate of salt uptake.
    (ii) Robertson and Turner (1945) have reported that respiratory inhibitors such as Azides, Cynamides etc., inhibit salt absorption.
    (iii) Lundegardh (1955,58) observed that increase in salt absorption results in salt respiration, indicating the involvement respiratory cytochrome system.

The Carrier Concept : This is the most widely accepted theory to account for the active migration of ions across the cell membrane. Basically, according to carrier concept, there will be certain special molecules called carriers which are energy rich. These molecules combine with the ions forming the ion carrier complex. The ion carrier complex travels across the membrane and in the cell interior the complex breaks releasing the ions. The carrier gets back to the membrane surface to receive more ions.

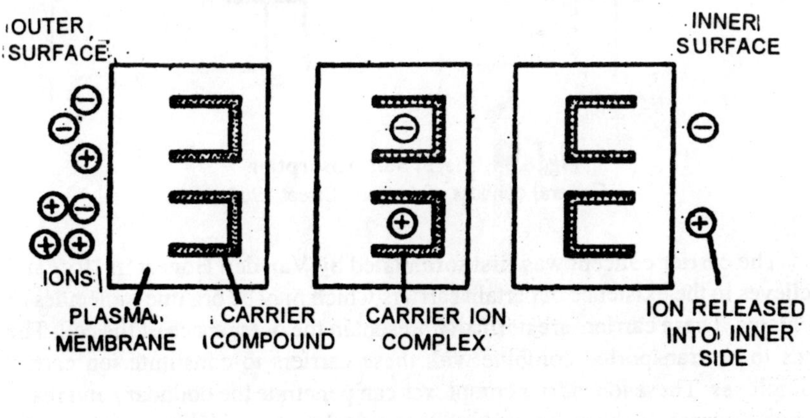

Fig. 8.4 Mineral Salt Absorption
Model of carrier concept

The whole volume of the cell can be divided into three zones. These are the outer free space, inner spac'e and the intermediate space. The outer space allows free diffusion of ions, while the inner space allows the entry of ions driven by the force of metabolici energy. It is believed that the intermediate spaces between the outer and the inner space and is impermeable to free entry by the ions. However the intermediate space is not clearly defined.

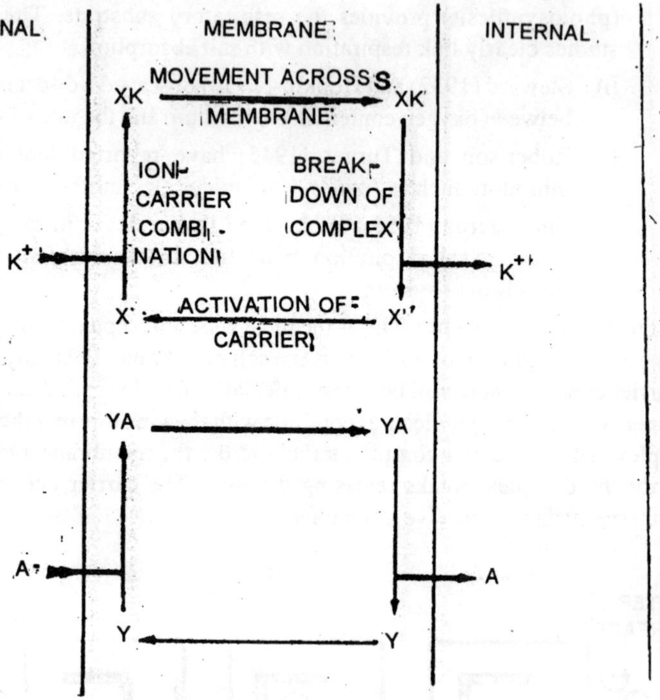

Fig. 8.5 Mineral Salt Absorption
General concept of carriers (Street 1962)

The carrier concept was first formulated by Van den Honest in 1937 and believes in the existence of certain carriers which may be organic molecules or vehicles. These carriers are normally present in the outer space of the cell. The ions to be transported combine with these carriers to constitute ion carrier complexes. These ion carrier complexes can penetrate the boundary and reach the cell interior. In the cell interior, the complex breaks up releasing the ion. The carrier comes back to the membrane outerspace to guide one more ion into the cell-interior. The ion carrier complex can move only in one direction ie., towards the cell interior. The ions released in the cell interior cannot come out and they accumulate. This theory explains specific absorption and abundance of particular ions in the cell interior, based on the affinity of the ions to the carrier molecules.

According to Street (1962) there are atleast two carrier molecules (one for cation and one for anion) which can be designated X and Y. These arise from precursors X^1 and Y^1.

These precursors obtain energy through the metabolic pathway (Respiration). The energised carriers *are,* present at the membrane outerface. They combine with ions K^+ and A^-. X combines with K^+ to form XK^+ and Y combines with A^- to form YA^-. These complexes cross the membrane and in the cell interior break up to release the ions K^+ and A^-. The carriers (X and Y) which are deenergised return to their precursor stage X^1 and Y^1,during their passage to the membrane outerface these precursors are reenergised (activated) and are now ready to accept fresh ions for transportation. The activation of the carrier is by the metabolic energy.

It is not certain as to how many carriers are there- whether only two or many. Some argue for only two carriers, one for all cations and one for all anions, and the point that needs clarification is the capacity of the carrier at a time- whether it carries only one ion or more than one ion. In other words how many active sites are therein a carrier where ions can combine.

The evidences for the existence of carriers comes from the following three phenomena- Isotopic exchange. Saturation effects and Specificity.

Isotopic Exchange

The entry of the ions into the cell takes place at two stages. At the first stage, they diffuse into the free space and can readily come out. At the second stage, ions diffuse actively into the inner space in a unidirectional way. The ions absorbed passively and held in the outer space can come out of the cell easily while the ions actively absorbed cannot be **deabsorbed.**

This shows that at least some quantity of salts are absorbed actively. This can be ascertined by using radioactive ions.

Using labelled ions several investigators have shown that, the membrane is impermeable to certain ions either from exterior to interior or *vice versa.* Legget and Epstein (1956) have conducted experiments on barley roots using labelled sulphate ions (S^{35}). They have found out that after some period of absorption of labelled sulphate, when barley roots are transferred to ordinary water, only some quantity of ions come out, while the rest is held back in the cell. The ions that are held back in the cell are actively transported hence remain in the inner space of the cell and cannot come out. These ions are also not available, for ionic exchange. This clearly shows that the entry and exit of atleast some quantity of ions with reference to the cell is mediated by carriers.

Saturation effects

Initially, increase in the concentration of salts in the external medium might enhance the rate of absorbtion due to increased availability. But this is subject to a certain limit. Any further increase of salt concentration beyond this limit will not influence the rate of salt absorption.This is due to the fact of fixed number of carriers. Initially all the active sites of carrier molecules will not be engaged, so that increase in salt concentration will engage all the active sites and hence increased absorption. But, after all the active sites are occupied any increase in salt concentration will not have any influence on the rate of salt uptake as all the carriers are saturated with ions. This is called saturation effect and proves the existence of carriers.

Specificity

Differential rate of absorption of different ions and varying concentration of salts in the cell interior are clearly indicative of selective absorption of ions mediated by carriers. The differential absorption is explained on the basis of specificity of certain carriers only to certain ions. Epstein and Hagen (1952) have shown that potassium, cesium and rubidium are carried by the same carriers while sodium and lithium have a different carrier. Similarly Legett (1956) has shown that selenate and phosphate compete for the same carrier, while phosphate and nitrate are not influenced by the concentration of selenate in the medium.

The specificity of ion and carrier is similar to enzyme substrate specificity (Devlin,1956). There are four important hypotheslis .regarding carriers. These are explained below.

I Protein Lecithin carriers

This theory proposed by Bennet Clark (1956) believes that certain amphoteric compounds act as carriers for both anions and cations. Such amphoteric compounds could be proteins associated with phosphatides (Lecithin). Lecithin is synthesized and hydrolysed in a cyclic manner; the cycle being called phosphatide cycle. Lecithin is synthesized on the outer surface and during its synthesis, it picks up an ion forming the ion carrier complex. The complex is hydrolysed in the cell interior releasing the ion.

The phosphate group in the lecithin acts as the binding site for cations, while the choline serves as the binding site for anions. In the cell interior, lecithin decomposes into phosphatidic acid and choline and at the same time releases the anion and cation held by it. Lecithin is regenerated from phosphatidic acid and choline by the enzyme choline acetylase and choline esterase involving the participation of ATP

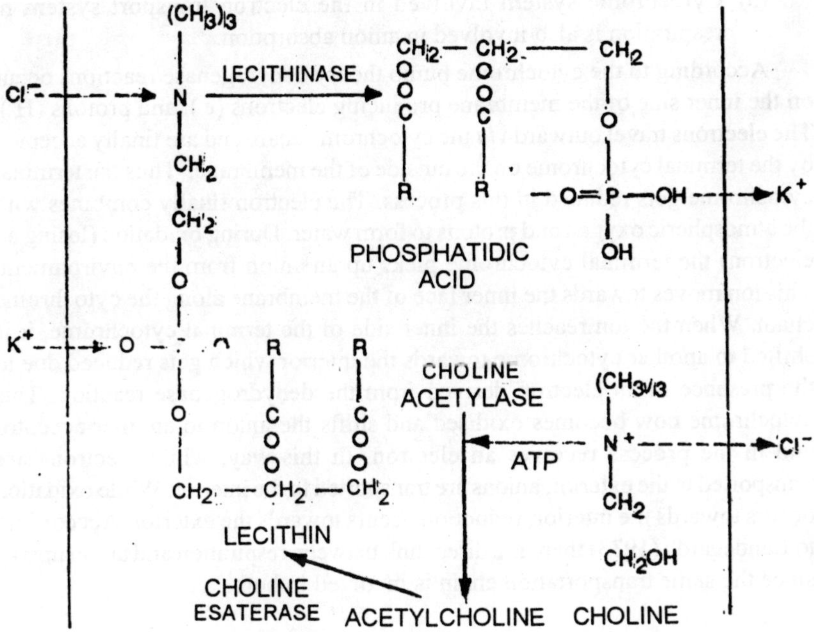

Fig. 8.6 Mineral Salt Absorption
Bennet Clark's Protein lecithin carrier theory

II Cytochrome pump hypothesis

This theory is based on the existence of a direct relationship between respiration and salt uptake. According to Lundegardh and Burstrom (1933), a quantitative relationship exists between respiration and anion absorption; such a relationship however does not exist between respiration and cation absorption. They observed that when a plant is transfered from water to salt solution, the rate of respiration increases. This increased quantity of respiration is called salt respiration. Lundegardh and Burstorm's theory is essentially based on the following points.

(a) Anion and cation absorption is independent and unrelated.

(b) Only anions are actively carried across the membrane.

(c) An oxygen concentration gradient exists from outer to inner surface of membrane in such a way that there is oxidation on the outer surface and reduction on the inner surface.

(d) Cytochrome system involved in the electron transport system in respiration is also involved in anion absorption.

According to the cytochrome pump theory dehyrogenase reactions occur on the inner side of the membrane producing electrons (e⁻). and protons (H). The electrons travel outward *via* the cytochrome chain and are finally accepted by the terminal cytochrome on the outside of the membrane. Thus the terminal cytochrome gets reduced in this process. The electron finally combines with the atmospheric oxygen and protons to form water. During oxidation (losing an electron) the terminal cytochrome picks up an anion from the environment. This ion moves towards the inner face of the membrane along the cytochrome chain. When the ion reaches the inner side of the terminal cytochrome, it is shifted to another cytochrome towards the interior which gets reduced due to the presence of an electron derived from the dehydrogenase reaction. This cytochrome now becomes oxidised and shifts the anion to another acceptor and in the process receives an electron. In this way, while electrons are transported to the exterior, anions are transported to the interior. While oxidation occurs towards the interior, reduction occurs towards the exterior. Acccording to Lundegardh (1933) there is a direct link between respiration and anion uptake since the same transportation chain is involved in both.

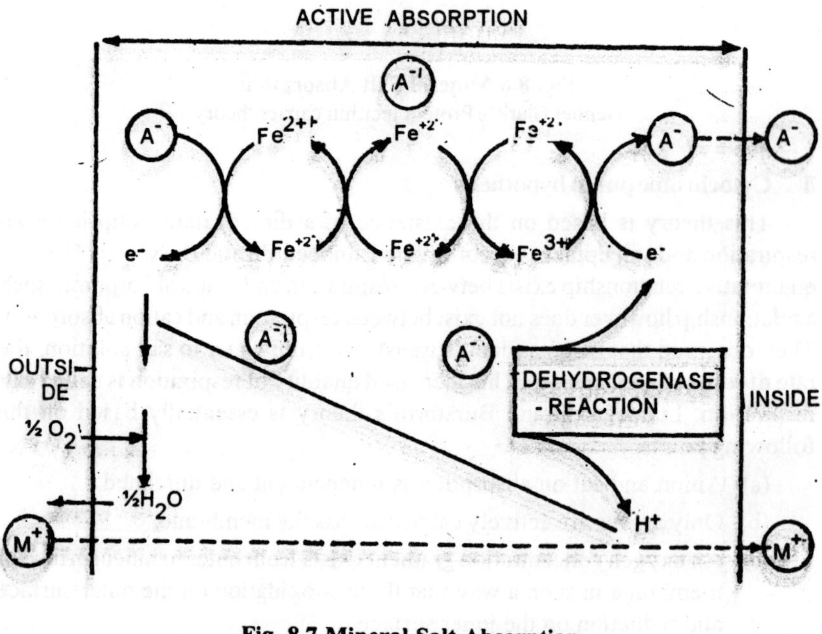

Fig. 8.7 Mineral Salt Absorption
Cytochrome pump hypothesis of Lundegardh

The theory however does not account for the absorption of cations. It simply assumes-that as anions are accumulated actively, cations move passively along the electrical gradient created by the accumulation of anions. The following are some of the evidences for the existence of cytochrome pump.

(a) Addition of salts to the medium in which living roots are kept will not only increase the rate of absorption of salts, but enhances the rate of oxidation of respiratory substrates as shown by spectrophotometric observations.

(b) Oxidase inhibitors have a similar mode of action on respiration and salt accumulation, thus indicating the involvement of cytochrome system in both the processes.

(c) Salt induced respiration is believed to be due to the increased outward flow of electrons which is brought about by rapid inward flow of anions.

Inspite of the fact that the cytochrome pump theory is quite appealing several objections have been raised against it. Some of these are

(i) There is carrier only for anions and not for cations

(ii) Involvement of metabolic energy is not explained properly in salt induced respiration.

(iii) Cations are also known to stimulate salt respiration (Hardley and Soverstreet 1955)

(iv) Selective ion uptake cannot be explained by the cytochrome pump theory.

(v) Robertson et al (1951) believe that the role of ATP is completely neglected in the hypothesis of Lundegardh. According to them 2,4 dinitrophenol (2,4 DNP) which inhibits only oxidative phosphorylation inhibits salt absorption also.

(vi) Salt absorption via cytochrome involvement cannot be explained in anaerobic forms as they (anaerobes) do not have a complex electron transport chain seen in aerobic forms.

(vii) If there is only one carrier tor all the anions, then competition should be there among them for binding sites. But as has been observed in several instances anions like sulphate, phosphate, nitrate etc., do not compete with one another.

III. Protein as carrier

According to Steward, a close co-ordination exists between protein synthesis and salt uptake. Steward believes that peptides of proteins can bind both cations and anions and release them in the cell interior. Sutcliffe (1962), another plant physiologist opined that proteins carry the ions up to the endoplasmic reticulum and release them. These ions combine with water in the cytoplasm to form vesicles which burst on the vacuolar surface releasing the ions into the vacuole.

According to Goldacre(1952), contractile proteins are involved in ion transportation. These proteins in their extended form bind to ions to satisfy the free valencies exposed to the membrane surface. When contracted the proteins release the ions to the cell interior, as the valencies are satisfied in the molecule in the folded condition. There is however no experimental evidence to this theory. Two models are proposed (Pardee,1968; Oseroff, 1973) to explain protein carriers. In the first one a protein binds to a solute (S) and rotates 180^0 within the membrane and releases the solute in the cell interior. In the second model the protein binds to the solute and undergoes conformational changes allowing the solute to be conducted through the pore.

IV Carriers Involving ATP (ATP theory)

The discovery of Robertson et al (1951) that 2.4 DNP an inhibitor of oxidation only, inhibits salt uptake supports the idea of involvement of ATP is salt transport. Laties (1957) and Sutcliffe (1959) also have proposed the ATP theories to account for active salt uptake. According to the ATP theory, hydrolysis of ATP molecules could energise the ion pump through the action of enzymes. Many of these enzymes have been found in the cell membrane. ATP plays a two way role in the cation upake ie., either by addition or removal of phosphate. At the first instance an organic compound is phosphorylated, which on dephosphorylation combines with a cation. The cation is released when phosphorylation takes place again.

In the second instance the phosphorylated organic compound combines with cation and the ion is released on the hydrolysis of the compound.

Raven (1967), working with the chlorophycean algae *Hydrodictyon* and *Nitella,* has shown that cation uptake is linked to ATP while anion uptake is related to the electron transport. The internodal cells in the alga *Nitella* are very big in size and have sodium extrusion pumps and inwardly pointed potassium and chloride pumps.

FACTORS AFFECTING SALT ABSORPTION

Salt absorption is affected by a number of external and internal factors. Some of the more important of the external and internal factors are- **External** a) temperature; b) pH; c) light; d) oxygen and Interaction; Internal a) growth and the physiological condition of the plant and b) aging. We shall discuss these briefly in the following paragraphs.

1. Temperature : Generally increase in temperature has an effect of increase on any physiological process. This is true for salt absorption also. But this is subject to the limit of optimum level. Any increase in temperature .beyond the optimum will retard the rate of salt absorption. Temperature effects are similar for both active and passive absorption. This is because of a similar relationship existing between free diffusion seen in passive absorption and temperature and biochemical reactions of active absorption and temperature. Extremes of temperature would kill the enzymes involved directly or indirectly in the process.

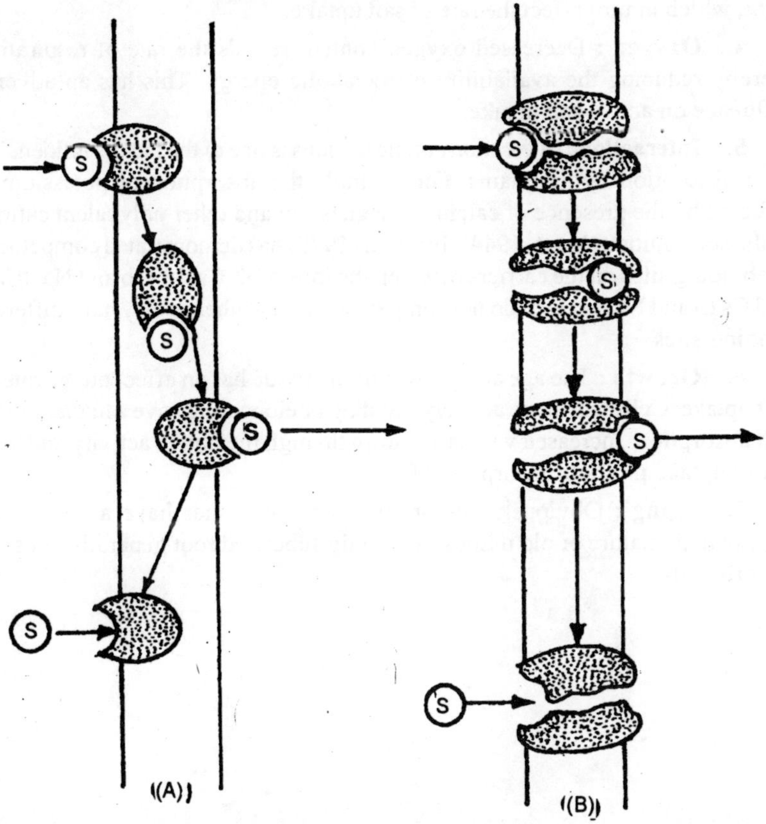

Fig. 8.8 Mineral Salt Absorption
Diagrammatic representation of membrane proteins acting as carriers

2. pH : changes in the hydrogen ion concentration level in soil solution affects the availability of ions to the plant. Generally decrease in the pH of soil solution enhances the uptake of anions. For example boron is absorbed rapidly as H_3BO_3 at a lower pH. Cations however are absorbed readily with the increase in pH. However pH values outside the tolerance limit of the plant will damage the root tissue and inhibit salt absorption.

3. Light : The effect of light on salt absorption is indirect. Other physiological processes like transpiration and photosynthesis which have a bearing on salt absorption are influenced by light. Gaseous exchange throught the stomata, synthesis of food during photosynthesis etc., are influenced by light, which in turn affect the rate of salt uptake.

4. Oxygen : Decreased oxygen content retards the rate of respiration thereby reducing the availability of metabolic energy. This has an adverse influence on active salt uptake.

5. Interaction : Interaction among the ions is one of the factors influencing their absorption by the plants. For example the absorption of potassium is affected by the presence of calcium, magnesium and other polyvalent cations in the soil solution (Viets, 1944). Eytstein (1978) has demonstrated competition for binding sites on the carriers between the ions of K, Cs, Li, Rb and Na. Rb,K and Cs Li and Na however do not compete with each other as they have different binding sites.

6. Growth : The age and growth of the tissue has an effect on the rate of salt uptake. Cells which are actively dividing or elongating have a higher rate of salt absorption. Increased vegetative growth, high metabolic activity and high water uptake promote absorption of salts.

7. Aging : Obviously young and growing tissues have a higher salt uptake than mature or old tissues. A heavily suberised root generally does not absorb salts.

9

TRANSLOCATION OF ORGANIC SOLUTES

In lower plants, with very little tissue differentiation, the site of food preparation is also the site of utilization.ln aquatic plants like algae, the absorption of inorganic nutrients and water is carried out by the entire general surface and the whole of the general surface again is involved in photosynthesis.ln higher plants however with complex tissue differentiation, the site of food preparation (leaves) and the site of its (food) utilization or storage are often separated far apart.Hence the organic food has to be transported to these areas. The phenomenon of food transportation from the site of its synthesis to the site of its utilization is known as Translocation of organic solutes.The term solutes is used, because food materials move (dissolved) in a solution. In the process of translocation, the movement of solutes from one cell to another which generally can occur in any part of the plantbody is not included.

The problem of the distance encountered by the migrating solute may not be much in herbaceous plants, but in the ca[illegible] of trees *like, Eucalyptus* and *Sequoia* it is indeed enormous because solutes many a time have to travel a distance of 300 - 350 feet.

Historically, studies on transolcation perhaps began as early as the middle of the 17th cent, even though the knowledge at that time was cursory. It was thought at that time that the 'elaborated sap' moved in the downward direction through the 'wood' to the roots.

Hartig (1837) was the first person who initiated studies on translocation. He gave the anatomical and physiological description of the tissues involved in translocation of organic compounds.He identified seive tubes in the 'bark' tissue and attributed a role to it in nutrient transportation.lt was he, who first demonstrated the collection of nutrients and the consequent bulging of the stem above a girdle (for details of girdling see later in the same chapter). Several scientists of 19th century after Hartig had engaged themselves in the study of translocation of organic soiutes.These include Sachs (1891) etc. Among the modern plant physiologists who have contributed a lot for the understanding of the translocation, the names of Schmidt (1914), Curtis and clark (1917-1950), Munch (1927), Masan and Maskel (1928), Esau (1966), Canny (1963), Fansan (1972) etc., are worth mentioning.

Rate of translocation : The rate and velocity of translocation varies with plants. Crafts and Lorenz (1971) measured the total amount of organic material in a mature pumpkin.Considering the approximate diametre of the stem tissue involved in translocation, and the concentration of the solute, they calculated that an average velocity of 11°cm/hour has to bemaintained during the deveiopment of the pumpkin fruit. Tracers injected into the stem during peak development showed a movement of 200cm/hr. In Soyabean, the rate of translocation may be a high as 1000-1500cm/hr (Nelson, quoted in Bidwel, plant physiology, 1979).

Direction of translocation : The movement of solutes in plants is basically of two types-Linear and Radial. The migration of the solvents in a vertical direction in the plants both upwards (in the case of germinating seeds and underground perenneating structures) and downward (from leaves to roots) is called linear translocation. Bulk of the translocation in plants is linear. Radial translocation, which is also common, though not same in quantity as linear, takes place during fruit and seed development.The direction here is perpendicular to the long axis of the stem.

Path of translocation : There is no unanimity among plant physiologists as to the actual tissue involved in translocation. However all are agreeable to the fact that the vascular tissues *viz.,* xylem and phloem are involved in the process.Even here it has to be pointed out that the overwhelming evidence is in favour of phloem, even though there are some evidences in favour of xylem also. Curtis and Clark and some others believe that phloem is the tissue involved in translocation, while Dixon and others believe xylem to be the path of translocation. We will examine some of these evidences.

Evidences for phloem : The following are the evidences in favour of phloem being the path of translocation.

(a) Elimination of other tissues : Though both xylem and phloem are ideally located for translocation, the former is not suitable atleast for downward translocation, because there is a rapid upward movement of water and inorganic salts in them. The cells of pith or ground tissue also may not be involved as sap concentration of transportable food in them is very less. Thus phloem is the only tissue suited for translocation.

(b) Ringing or Girdling : This involves the selective removal of a part of the stem tissue in the form of a ring or girdle. Two types of rings of tissues are removed; in one only the pholoem is removed and in the other only the xylem is removed. In the ringing experiment three plants are selected.

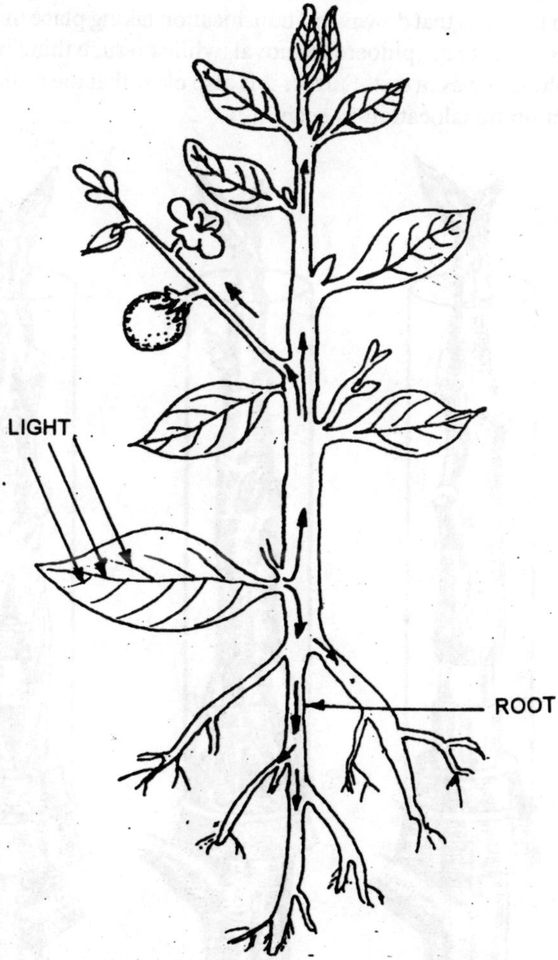

Fig. 9.1 Translocation of organic solutes
Path of movement of organic solutes within the plant

Plant A	only xylem is removed.
Plant B	only phloem is removed.
Plant C	control (no tissue is removed).

The plants are kept in sunlight and allowed to photosynthesize. After some time it will be noticed that in plant B the portion of the stem immediately above the ring bulges, while no such thing is observed in A or C. This is explained on the basis that downward translocation taking place through phloem is affected in B due to its (phloem) removal, while no such thing happened in A or C as the phloem was in tact. Further it is also clear that the removal of xylem has no effect on translocation as seen in A.

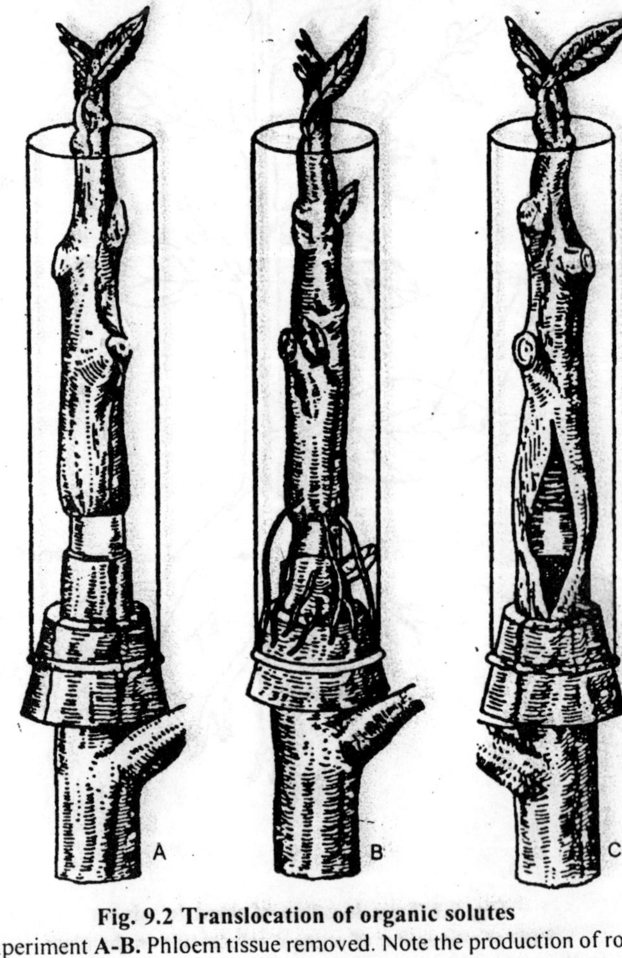

Fig. 9.2 Translocation of organic solutes

Ringing experiment **A-B.** Phloem tissue removed. Note the production of roots above the ring. **C.** Xylem tissue removed. (Control is not shown in the figure)

Similar ringing experiments maybe conducted to show that the upward translocation also is conducted through phloem. In this experiment also three sets of plants are taken A - Xylem removed, B - Phloem removed and C - control. In the plants A and B shoots are defoliated above the rings. The cut surface of the stem is enclosed in a glass cylinder filled with water to keep the tissues moist. Plant A in which only xylem is removed shows elongation of the stem indicating undisturbed movement of solutes through phloem. Plant B shows no elongation indicating stoppage of solute flow as phloem has been removed.

Structure and distribution of phloem : The phloem tissue is composed of sieve; tubes, companion cells and phloem parenchyma. The seive tubes are elongated like the vessels of xylem and are joined end to end forming a continuous system throughout the plant. The perforated seive plates allow for cytoplasmic connections to help in conducting the solutes. The presence of a second group of phloem as seen in bicollateral vascular bundles (cucurbitaceous plants) found in thin stems with large leaves and large sized sieve tubes seen in large fruits is an additional evidence for phloem being involved in translocation.

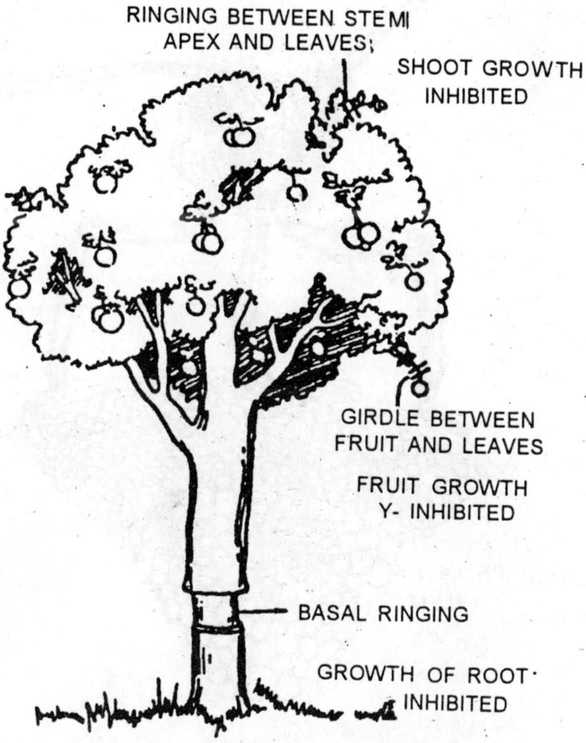

Fig. 9.3 Translocation of organic solutes

Effect of girdling in various parts of the plant.

Effect of blocking phloem : Occassionally the sieve pores are blocked by callose deposits resulting in stoppage of translocation.

Chemical analysis of phloem sap : Chemical analysis of the sap extracted from phloem reveals that there is a high concentration of soluble sugars.Since the carbohydrates are translocated in the form of sugars, it clearly indicates the phloem being the tissue for translocation. Further, the concentration of sugars seen in the sieve elements is very high compared to what is necessary for the maintenance of normal osmotic relationship with the adjacent cells.

Mason and Maskel (1928 - 34), in their experiments on cotton plants showed that variations in sugar and organic nitrogen content in the leaves were followed by similar variations in the bark, while the xylem had the concentrations of these substances constant.

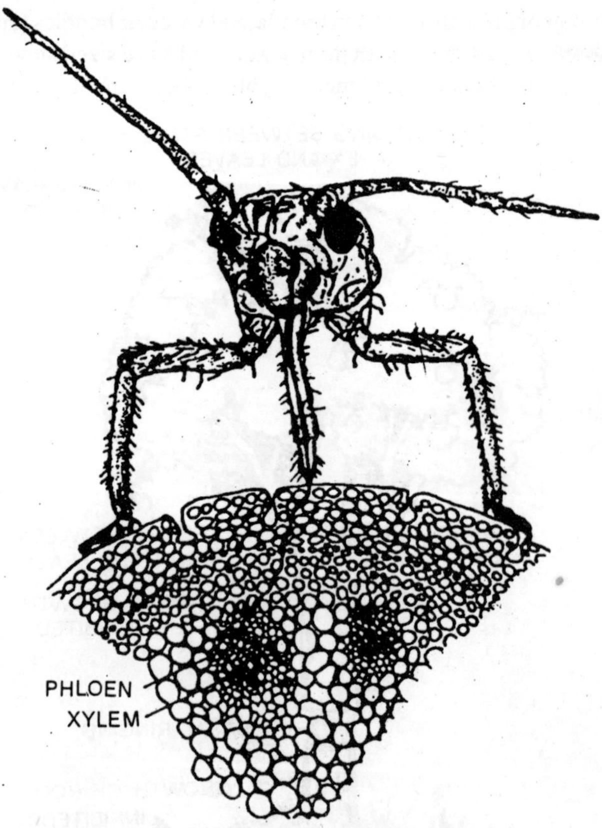

Fig. 9.4 Translocation of organic solutes
Feeding of an aphid. Note the penetration of stylet upto phloem

Kennedy and Mittler (1953) have developed a more refined technique of phloem sap analysis making use of stem feeding aphids. These insects inject their stylets into the stem and suck the phloem sap. If the stylet is cut, phloem sap continues to exude for several days indicating the presence of a positive pressure. Weatherlay and his colleagues have analysed this sap and have found sucrose to be the principal carbohydrate. Besides sucrose, sucrose based oligosaccharides are also found in the sap. Among the nitrogenous compounds, amino acids and the amides glutamine and Aspargine are important. In pea seedlings homoserine is the nitrogenous compound translocated through phloem.

Isotopic studies : Radio isotopes of many elements have been of great help for measuring velocities, rates, and even path of translocation. Rabidean and Burr (1945) supplied labelled carbon (C^{14}) to the plant and allowed it to photosynthesize. They found the bidirectional movement of the labelled substance. A section of the stem was killed subsequently and this stopped the movement of C . Since conduction can still take place in xylem even after the living tissues are killed, translocation should be through phloem. Bonner and Went (1944) arrived at the same conclusion after the isotopic experiments.

Two isotopic experiments are worth mentioning in this context. In the First one, conducted by Stout and Hoagland in the willow plant, the roots were supplied with K and the phloem and xylem were separated, for a short distance by inserting a wax paper. Results showed that potassium migrated in xylem and not in phloem. But when phloem and xylem were in contact as above and below the level of the insertion of paper, lateral conduction occured and potassium was found in phloem also.

In another experiment Biddulph and Markle, supplied labelled phosphate and found that it moved only in phloem and not in xylem.

Radioautographic techniques have conclusively proved the path of translocation to be phloem. In this method, tracers are supplied to the plant and thin sections of the stem are cut after sometime. These sections are placed on a slide on which is placed a thin photographic emulsion. The film is developed subsequently. Microscopic examination of the film reveals that labelled elements are accumulated in the regions of phloem.

Evidences for Xylem : According to many physiologists like Dixon etc., xylem is the path of movement for organic solutes also besides inorganic nutrients. The following are some of the evidences.

(a) Injection of dyes : Certain dyes injected into the plant are known to move both up and down in xylem. But it is difficult to visualize how there can be a downward migration when there is a positive pressure of the upward flow.

(b) Analysis of xylem sap : It is incorrect to say that the transpiration stream in vessels, contains only inorganic constituents. Often, as during spring in trees, the sap is rich in sugars produced by the hydrolysis of starch stored in the living cells of wood. It is beleived that sugars are transported up in the transpiration stream. There is also evidence to indicate that organic nitrogen compounds appear to move up in the xylem of trees, although they are transported through phloem in herbaceous plants.

The movement of sugars in the xylem is doubtful, although its presence cannot be doubted For, during spring when maxirtium translocation takes place, xylem vessels seem to have a low concentration of sugars.

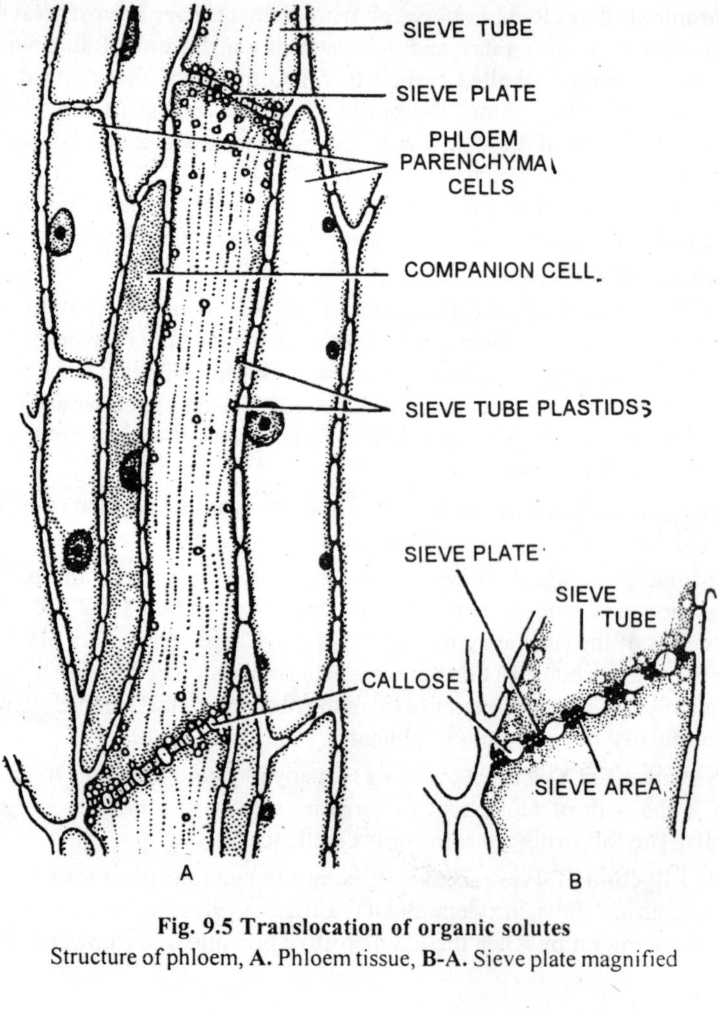

Fig. 9.5 Translocation of organic solutes
Structure of phloem, **A.** Phloem tissue, **B-A.** Sieve plate magnified

In conclusion, it may be mentioned that the final word on the path of translocalion has not been said. It is possible that phloem may not be the only vehicle for all organic solutes, though there are overwhelming evidences in favour of it.

Detailed structure of phloem : Phloem is a much more complex tissue than xylem. According to some plant physiologists, the exact structure of phloem is still a matter of conjecture (Bidwell 1979), because its fine structure is destroyed or altered during preparation for microscopic examination.

The phloem is mainly composed of *sieve tubes* and phloem parenchyma. In Angiosperms, the sieve elements are always accompanied by companion cells, while in gymnosperms albuminous cells replace companion cells. Companion cells are living cells and are nucleated. The sieve elements however are enucleated, but unlike the vessel elements of xylem the sieve elements have protoplasm. They can be plasmolysed indicating that they are bound by a selectively permeable membrane, but they do not have a tonoplast. Much of the protoplasm present in sieve tubes is in the form of **P-protein,** a fibrillar protein. The exact structure and function of this P-protein is still to be unravelled completely (Bidwell 1979). According to Esau, in mature sieve tubes, the cytoplasm and vacuole became a single system called **myctoplasm.** Ribosomes and dictyosomes are absent. Mitochondria however are present, but they differ from the normal ones in having less cristae.

During the course of maturation, the cell wall, specially the transverse wall undergoes changes. The cross wall becomes porous, and through each pore runs a cytoplasmic fibril connecting the protoplast of another sieve tube. Occasionally the lateral walls also may be porous. The end walls, where the pores are many in number are called sieve plates. Occasionally the sieve pores are plugged with proteinaceous masses called **slime bodies.** In some sieve tubes however, the pores are partially blocked by **callose** (carbohydrate) deposits. It is generally beleived that in an actively translocating sieve tube the pores are largerly unplugged.

The phloem consists of phloem fibres, ray cells, sclerids etc; besides seive tubes and companion cells. These other cells however do not seem to play any role in translocation. Companion cells however are ontogenetically and physiologically related to the sieve elements.

The phloem mother cells cut off one or more companion cells before they mature into sieve tubes. Walls separating the two are thin and highly pitted. Death of companion cells renders the sieve tubes functionless. Since the seive elements are enucleated it has been suggested that the companies cells have nuclear control over seive elements (Esau, 1939). According to Bieleski (1966), the companion cell and the seive tube should be viewed as a single functional unit.

GENERAL ASPECTS OF PHLOEM TRANSLOCATION : The following are some of the aspects of phloem translocation.

(a) Bidirectional movement : The organic solutes move both in upward and downward directions in the seive tubes. In an experiment using labelled carbon, Biddulph and Cory (1957) have shown that the leaves near the base of the stem transport food to the roots, while those near the stem apex transport the food up. The leaves in the middle, transport food in both directions. It is however not certain, whether there is a unidirectional movement in a given seive tube in a given time or the same seive tube can conduct simultaneous bidirectional translocation. Biddulph and Cory (1953), however have demonstrated that separate phloem bundles are involved in bidirectional translocation.

Exthange between phloem and xylem : Evidence exists to prove that materials get exchanged between xylem and phloem during translocation and the ray tissues acts as a channel for this. Webb and Gorham (1965) have reported the radial movement of labelled (C^{14}) photosynthate in stems, petioles and hypocotyls of squash. It is but natural to expect a radial migration, since phloem is the principal tissue involved in the supply of organic food to all the non green cells. According to Porter (1966) about 12% translocation is radial. Zimmermann (1961) has estimated 1 cm of lateral flow for every 50 cm of linear flow of solutes.

Vein loading or phloem loading : Substances that migrate from on cell to another have to cross many membrane barriers. Such a transport is called **transcellular transport.** Phloem loading or vein loading is one such example of transcellular transport. Sugars produced in the photosynthetic cells move across several parenchyma cells before they reach the sieve elements of the vein in the leaf. This movement of solutes from the sites of photosynthesis into the seive elements is called **vein loading or phloem loading.**

Sucrose is the form in which the photosynthate is translocated. To begin with, sugar is transported across the plasma membrane of a photosynthetic cell into the water in the apoplast (cell wall). From there, the sugar migrates from one apoplast to the other and finally makes it to the phloem. Finally the sugar crosses the barrier of the plasma membrane of the sieve element. The migration of sugar from the apoplast of a parenchyma cell into the sieve element is called phloem loading. The transport of sugar across the plasma membrane of the sieve element is a sucrose-proton co-transport process and is discussed below.

A model for sucrose proton co transport : The plasma membrane contains the enzyme ATPase, which hydrolyses a molecule of ATP and simultaneously pumps two protons out of the cell. Thus the outside of the membrane becomes positive and the inside negative. As a result a charge imbalance is created across the membrane. Also present on the outside of the membrane is a sucrose proton translocator. When the translocator is negatively charged, it accepts a molecule of sucrose and proton. This complex migrates across the membrane mainly as a result of electrical imbalance (proton motive force). Sucrose and proton are released in the cell interior. The tranlocator restored to its original form is ready to react with one more complex of proton-sucrose on the membrane exterior. Sucrose will continue to accumulate inside the cell, until the difference in the chemical potential of proton on two sides of the membrane equals the difference in chemical potential of sucrose on either side of the membrane.

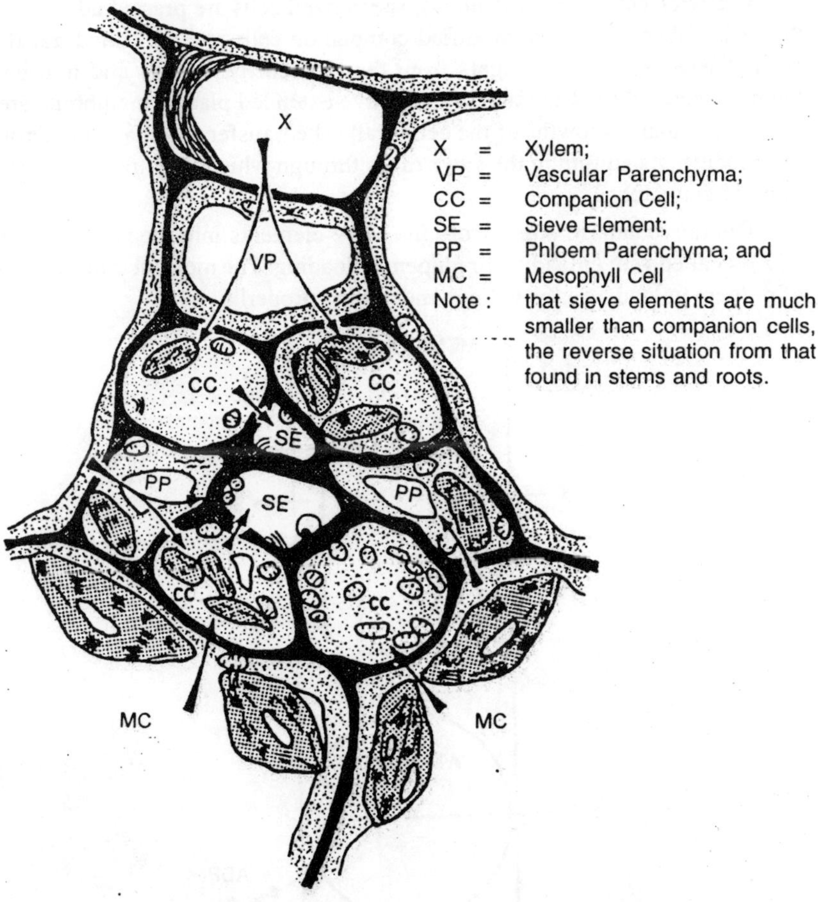

Fig. 9.6 Translocation of organic solutes

Cross section of a veinlet of leaf to show the entry route of solutes into phloem (Diagrammatic representation of electron microscopic view (after Giaquinta 1983)

The pumping of protons across the membrane (creating the proton motive force) as a result of ATPase action is the basic energy source, not only for sucrose transport, but for all other 'active transports'.

There are several translocators in the plant system. The sucrose proton translocator is called a symport. A symport helps in the movement of linked entities simultaneously in the same direction. Two other types of translocators found in higher plants are uniport and antiport. A uniport helps in the movement of a single entity across a membrane at one site. Antiport helps in the linked migration of two entities in opposite directions at the same site.

Transfer cells : In some plants, specialized cells are present adjacent to the seive tubes. These are modified companion cells and are called transfer cells. These cells absorb sugars from the chlorenchyma cells and transport them to seive tubes. The transfer cells have extended plasma membrane area due to irregular ingrowths of the cells wall. The transfer cells have numerous plasmodesmata adjoining the sieve tubes through which they transport sugar into the sieve tubes.

The movement of sugar from the seive elements into the apoplast (cell wall) is called vein unloading or phloem unloading. The molecular mechanism of phloem unloading is as yet not understood properly.

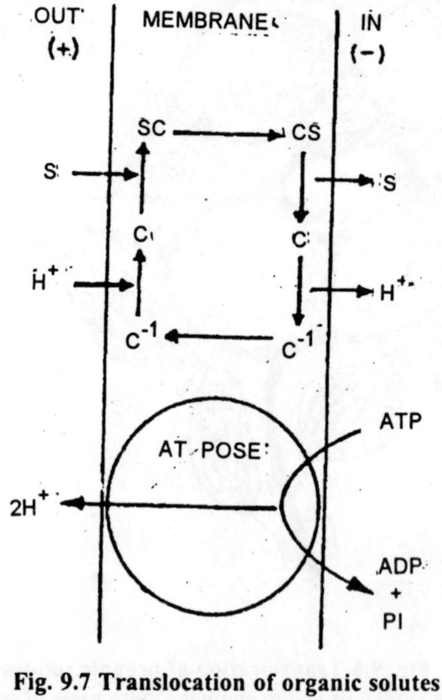

Fig. 9.7 Translocation of organic solutes

A model of Sucrose proton transport. (After Noggle and Fritz 1983)

S-Sucrose, C-Translocator, SC and CS Sucrose proton - translocator complex

Translocation in seive tubes : Many of the evidences favour the flow of organic nutrients in the seive tubes. This movement may be both downward and upward depending on the situation in the plant. This translocation in seive tubes is often called long distance transport. In discussing translocation it has become customary to refer to the photosynthesising cells as the sources, the seive tubes as the channel and the regions that utilize the sugars as the sink. Usually the concentration of the sugars is always more in the source than in the sink, hence the nutrient flow is along a concentration gradient from source to the sink.

Even though there is overwhelming evidence for phloem translocation and majority of plant physiologists favour this, there is however no unanimity regarding the mechanism (forces responsible) of nutrient flow in seive tubes. There are several views and the peculiarity seems to be, there are sufficient evidences for some of the views at least. It is quite possible that nutrient flow in phloem is driven by more than one mechanism or different mechanisms might operate in different plants or in different tissues of the same plant or at different times in one and the same tissue.

MECHANISM OF TRANSLOCATION : The following are the major theories to account for the mechanism of phloem transport of nutrients.

1. Diffusion hypothesis.
2. Activated diffusion and pumping.
3. Protoplasmic streaming.
4. Transcellular streaming.
5. Interface diffusion.
6. Electro osmosis.
7. Contractile proteins.
8. Mass flow (Bulkflow) hypothesis.
9. Modified pressure flow theories.
10. Biomodial theory.

Diffusion hypothesis : Many plant physiologists are of the opinion that nutrient translocation takes place following the principles of diffusion. Since nutrient flow is from the region of higher concentration (source) to the region of lower concentration (sink) where it is either utilized or converted into insoluble form, it must be according to the process of diffusion. Even though the theory seems to be plausible on the face of it, there are several strong objections. One is that of the rate of flow. If we consider the translocation as a simple process of diffusion, the rate of flow must be according to the concentration gradient. But the observed rate of flow of nutrients in the sieve tubes is atleast 40,000 times more than what is possible by the process of diffusion. This rate of flow is surely not dependent on concentration gradient. Secondly, respiratory poisons affect the rate of flow indicating its dependence on metabolism.

2. Activated diffusion : According to British physiologists Mason and Maskel (1928) and Mason and Phillis (1936), the solute transport in the seive tubes, (though essentially a process of diffusion) is an activated process. The normal rate of diffusion is greatly enhanced due to the protoplasmic properties of seive tube, which reduce resistance to the flow of nutrients. According to them, the activation of diffusion, increasing the flow rate is due to reduced resistance of cytoplasm.

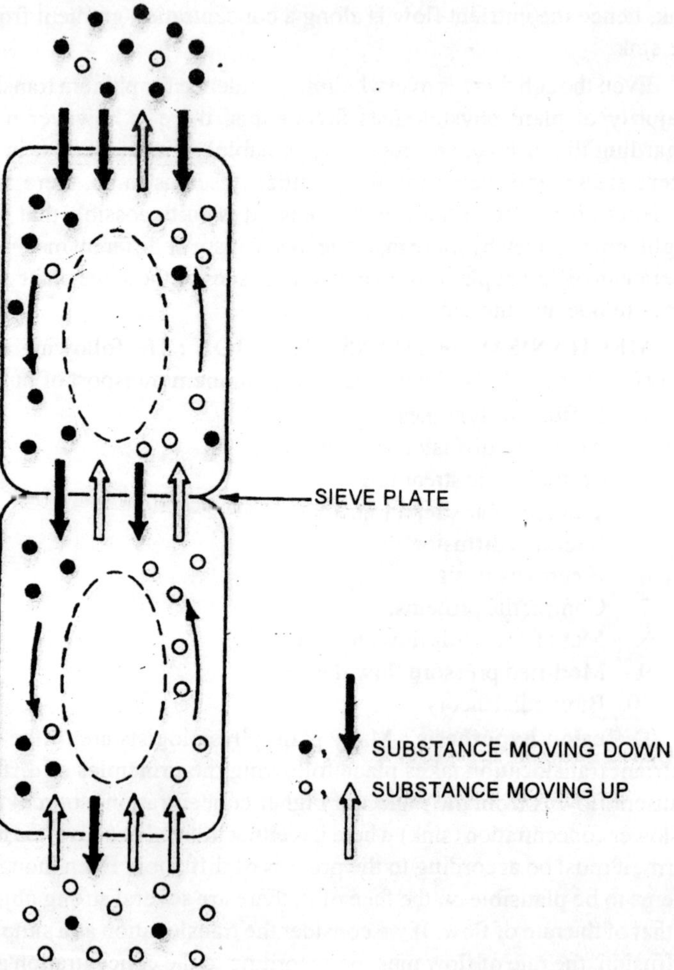

Fig. 9.8 Translocation of organic solutes
Protopiasmic streaming bringing about stimultaneous two way transport

Various pumping mechanisms have been proposed as existing in the seive elements. According to one, many pumping 'points' exist in the seive element and these are powered by the metabolism of either the companion cell or the seive element itself. But there is no experimental evidence for this.

3. Protoplasmic streaming : This is one of the oldest theories proposed for the mechanism of translocation. Proposed chiefly by the Dutch Botanist Hugo de Vires, the theory suggests that protoplasmic streaming (cyclosis) observed in several cells including sieve elements is responsible for the translocation of organic solutes. After reaching the end of a seive tube, the solutes migrate to the next seive tube by diffusion. Thus solute translocation is a combination of diffusion and protoplasmic streaming.

According to some plant physiologists protoplasmisc streaming is continuous from one sieve element to the other through the sieve pores In such a case atleast two substances can move in the sieve elements in opposite directions, depending on their own concentration gradient O.F.Curtis (1935, 1975) an American plant physiologist is a strong supporter of this theory.

One of the evidences to support the protoplasmic streaming theory, comes from the fact that cyclosis in seive elements is affected by factors that influence metabolism and translocation indicating a connection between these. Very high rates of cyclosis have been observed insome seive elements (Bidwell R.G. 1979).

According to Esau (1963) however, there is no cyclosis in seive tubes. It has also been observed that cyclosis even if it exists in young seive elements, decreases with maturity.

Another objection against cytoplasmic streaming being responsible for solute movement comes from the fact that, the nutrient rich sap coming out of cut seive elements is from the vacuole and not from the cytoplasm. This indicates that the solutes are transported through the vacuole of seive tubes. According to Esau (1966) however the tonoplast disappears in mature seive tubes resulting in the 'unification' of cytoplasm and vacuole into what is called *'mytoplasm'*. According to Bidwell (1979), on energetic grounds also cytoplasmic streaming theory can not be accepted.

4. Transcellular streaming : Thaine (1961, 62, 64) and Canny (1962) are of the opinion, that sieve elements have a number of cytoplasmic strands called **transcellular strands** traversing along their path. They have observed movement of solutes in opposite directions in adjacent strands. The existence of transcellular strands can conveniently explain bidirectional movement of solutes in one and the same seive element. According to Thaine, the transcellular strands in sieve elements possess mitochondria like particles, which provide energy rich ATP for the movement of solutes. Radio isotopic experiments conducted by Thaine usipg C^{14} have indicated the existence of radioactive strands within the seive elements. Thaine, further proposed that transcellular strands are poteinaceous and have microtubules helping in rhythmic contraction.

Aikman and Anderson (1971), who supported Thaine assumed that peristalic contractions occur as series of waves along the length of transcellular strands when there is movement of the solutes.

The theory of transcellular streaming is very appealing as it explains easily the bidirectional flow of solutes.

Several objections have been raised against this theory. These are:-

(a) How biological energy is transduced into a physical force in a sieve tube?

(b) What is the nature of the transcellular strands? Are they solid or tubular?

(c) Whether the solute particles move in or on the strands.

(d) Why the transcellular strands are not seen when sieve tubes are observed under electron microscope?

5. Interface Diffusion : Any substance that has the capacity to lower surface tension at the boundary between two immiscible liquids or between a liquid and gas get rapidly distributed across the entire boundary. This is a phenomenon normally observed, when gasoline spreads rapidly across the surface of water. The rate of diffusion at such boundaries is found to be 50,000 times more than the normal rate of diffusion. Many plant physiologists are of the opinion that the cytoplasm of phloem elements contains such a boundary system, and the solutes migrate at rapid rates due to interface diffusion. R.H.Vanden Honert, a Dutch physiologist who suggested this hypothesis, is of the opinion that diffusion gradients are maintained due to rapid addition and removal of the photosynthates (solute) from the system. But many scientists doubt whether the cytoplasm of sieve elements offers sufficient boundary area for the occurrence of interface diffusion.

6. Electro osmosis : Diffusion of water and salts across a membrane subjected to different electrical potentials is called *electro osmosis.* The migration is mainly due to electrical gradient. According to the electro osmosis theory proposed by Fenson (1957) and Spanner (1958), the sieve tube membrane is porous with fixed charges and the electrical potential across the sieve plates brings about the migration of solutes. Active transport of K^+ into the sieve plates and the secretion of the same (K^+ ions) on the other side of the plate produces circulatory pathway. K^+ ions are pumped into the sieve tubes at the expense of ATP. Sugar molecules adhered to K^+ ions are carried across the sieve plate, thus sugar molecules migrate across the sieve plate coupled with

K^+ ions which generate an electrical gradient due to their circular movement between companion cells and sieve tubes. One of the experimental evidences for this theory comes from the work of Bowling (1968) who reported the existence of electrical potential difference on either side of the seive plate in the primary phloem of *Vitis vinifera.* The observed difference of 4-48mv was found to be sufficient for the electro osmosis flow of nutrients. Electroosmotic theory does. not account for the bidirectional movement of nutrients.

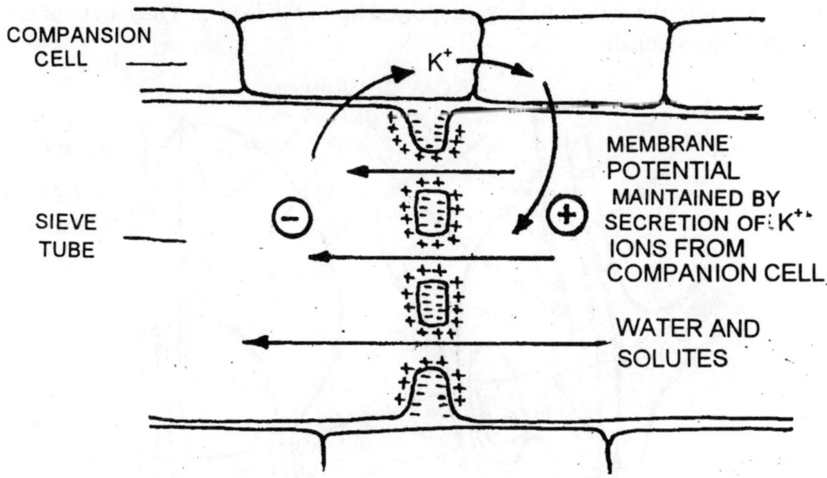

Fig. 9.9 Translocation of organic solutes
Electro-Osmotic theory of Spanner (A diagrammatic representation)

7. Contractile proteins : Fenson and William (1974) have reported the presence of a reticulum of microfibrils in the sieve tubes. According to them, these microfibrils oscillate and bring about the solute migration within a sieve tube. The microfibrils are composed of P-proteins and their oscillation resembles that of flagella seen in some microorganisms. There are neither strong evidences nor objections to this theory.

8. Mass flow hypothesis : Mass flow or bulk flow of solutes in sieve elements was first proposed by the German Scientist Munch (1931). Munch visualises a pumping action in mesophyll cells. According to Munch's hypothesis, the sieve tubes are connected to one another by means of cytoplasmic connections, forming a continuous system called symplast. The symplast is impermeable on the outer face.

Mass flaw of solutes as per this hypothesis is as follows. A very high concentration of nutrients is present in the mesophyll cells due to two reasons

- a) water is transpired continuously from the mesophyll cells and b) food is continuously synthesised. Solutes are actively loaded into the phloem thus increasing their osmotic potential. This increased potential draws water from the xylem vessels by means of osmosis, and hydrostatic pressure is built up in the sieve tubes. In the cells of the root, food material is continuously utilized reducing their osmotic potential as well as turgor pressure. Thus a turgor pressure gradient is established resulting in the mass flow of solution from supply end (leaf) to consumption end (root). At the consumption end water diffuses into the xylem to be transported up to the leaves. Thus a circulatory system is established.

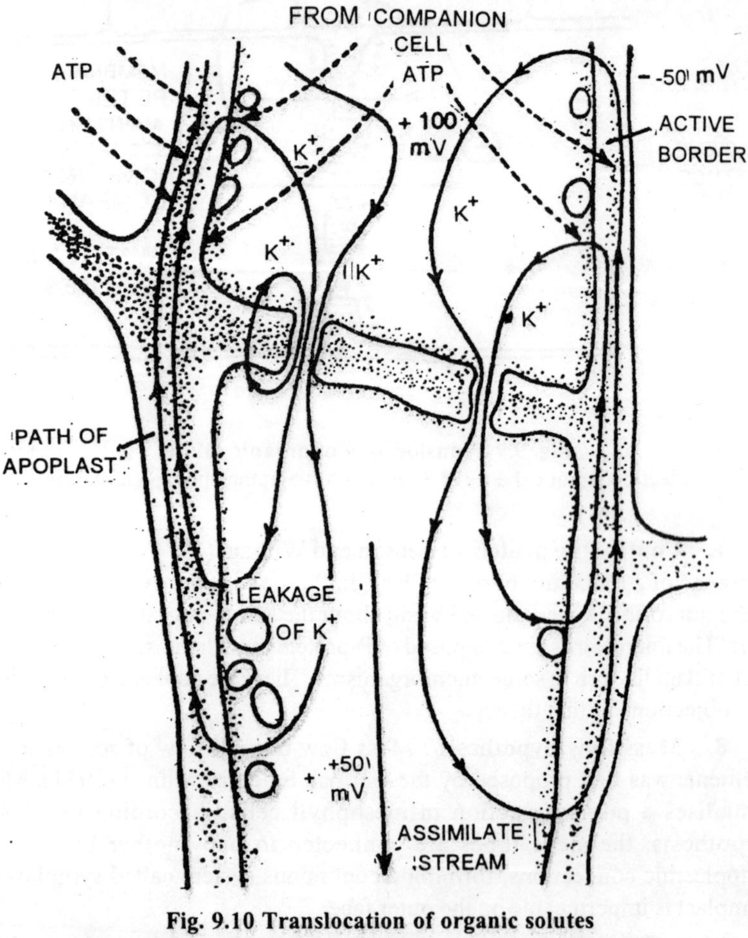

Fig. 9.10 Translocation of organic solutes
Diagrammatic long section of Phloem elements to illustrate the electro-osmotic theory

The principle of mass flow may be illustrated with the help of a simple experiment. Two chambers A and B with selective permeability are connected by a tube T containing water. Here chamber A is comparable to mesophyll cells, chamber B to the root cells. .Tube T is comparable to phloem. This forms a closed system similar to what is seen in plants.

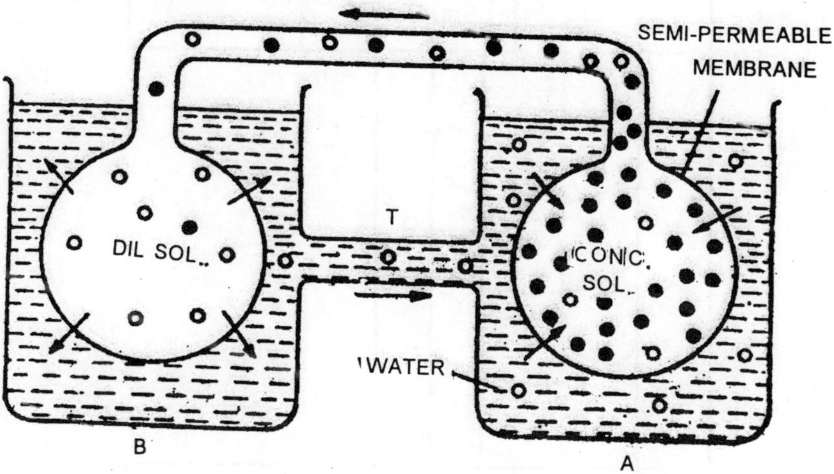

Fig. 9.11 Translocation of organic solutes
Model to explain Munch's mass flow hypothesis

Chamber A contains a highly concentrated sugar solution.while in chamber B its concentration is low. The system is dipped in a water filled vessel. In this experimental set up A corresponds to supply end, B to consumption end and T to longitudinal sieve tube system. The outer vessel having water corresponds to xylem. The concentration at A causes rapid diffusion of water into it resulting in a high turgor pressure. The solution from A therefore flows *en masse* to B until turgor pressure gradient is maintained. The flow stops, when turgor pressure both in A and B are equal. However if a high turgor pressure is maintained at A, there is a continuous flow from A to B. This is possible in the plant by the continuous synthesis of food. On reaching chamber B the water will flow into the outer vessel (xylem).

The main aspect of mass flow is that it requires a positive hydrostatic pressure in the seive tubes and a continuous supply of sugars in the leaves to generate enough pressure. Experimental evidences have clearly pointed to the presence of a positive pressure flow in phloem. An evidence in favour of this theory is that growth substances or viral particles are transported rapidly in sieve elements, when the leaves are illumminated. This suggests that sugars produced during photosynthesis will help generate the pressure gradient.

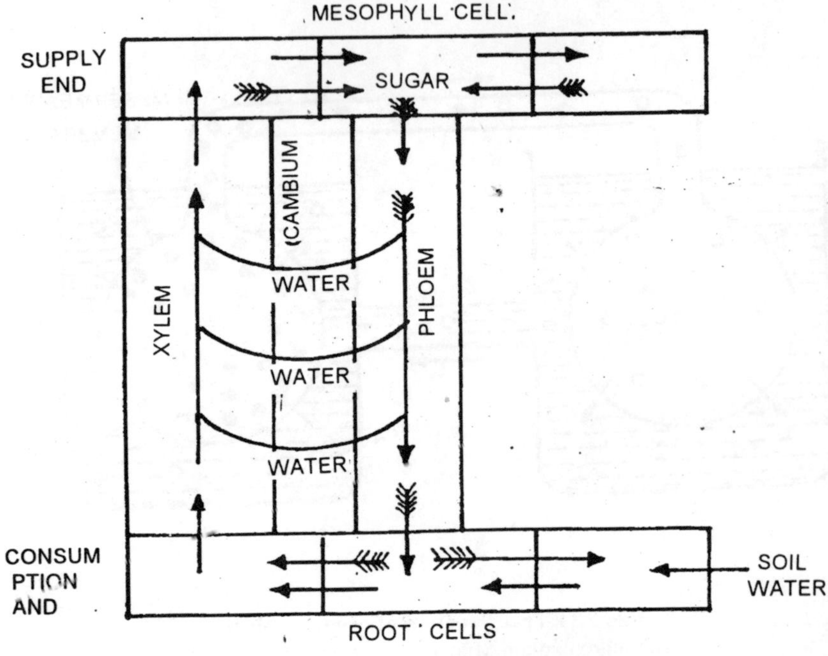

Fig. 9.12 Translocation of organic solutes
Diagram to explain flow of solutes within the plant as per Munch's hypothesis

There are several objections to mass flow hypothesis. These are-

(a) Mass flow is possible only when the sieve pores are not plugged. But this is not always the case.

(b) Mass flow can account for only a unidirectional movement. It is inconsistent with simultaneous bidirectional movement. But one need not always assume that bidirectional movement should take place in the same place in the same sieve tube. It is quite possible that different sieve tubes are engaged in translocation in opposite directions simultaneously. This is compatible with mass flow (Crafts 1951). One of the recent experiments (conducted) using C^{14} has shown that there can be bidirectional movement in one and the same sieve tube.

(c) Another objection against mass flow is that the observed osmotic pressure (OP) at the supply end and consumption end does not always assure a positive gradient. Many a time O.P of receiving cells is more than that of the supply cells.

(d) Studies on translocation on the effect of low temperature on sieve elements have shown that the sieve element cytoplasm actively participates in the proces, while mass flow assumes them to be only conduits for translocation flow. Kursanov and Brovchenk (1961) have shown that translocation is triggered on the application of ATP. It is also widely held that even the diffusion of sugars from sieve elements to the receiving cells is an active process.

9. Modified pressure flow : Three modifications have been proposed to the mass flow or pressure flow hypothesis. These are - volume flow, solution flow and active mass flow.

According to the **volume flow** hypothesis proposed by Eschrich, Evert and Young (1972), the pressure difference which acts as the force for the solute flow is not an overall one. It is more localized. According to them hydrostatic pressure differences across the sieve plates are responsible for solute flow.

Solution flow theory proposed by Marshall and Wardlaw (1973) assumes that both the solute and the solvent move concurrently. Further they assumed that different solutes move at the same speed as they are driven by physical forces.

In the **active mass flow** theory proposed by Mason and Phyllis (1936), the cytoplasm of sieve tubes requires oxygen to maintain its active state and to bring about translocation. They have shown that in living sieve tubes the movement is 40,000 times greater than in a corresponding column of water in a glass tube.

10. Bimodal theory : Fenson (1971) has proposed his bimodal theory of translocation, which in a way is a combination of mass flow and activated diffusion. According to him, sucrose is translocated in two modes. The first mode of translocation is brought about by the microperistalic (pumping) movement along the hollow fibrils of contractile proteins present in the sieve tubes. The second mode of flow is mass flow around the contractile proteins. According to Fenson, pulsating movements of proteins may help move sucrose at the rate of 400cm/hr. This method of translocation is not hindered even by callose plugs because, protein fibrils strech between successive sieve tubes. It has been estimated that an average of 50-350 microfibrillar proteins can pass through a single sieve pore.

Factors influencing translocation : A number of environmental factors affect the rate of translocation. Some of these are:-

1. Temperature : According to Swanson and Whitney (1953) translocation is intimately related to temperature variations. They observed a translocation peak at 25 - 30^{0}C. Lower ranges of temperature generally have a retarding effect on translocation.

2. Light : The effect of light on translocation is indirect, though it is very profound. Hart et al (1964) have observed in sugarcane plants, that translocation is reduced to half its normal rate in the absence of light. Low intensities of light inhibit root and shoot growth resulting in lesser demand of nutrients from these parts. Hart et al (1964) have found that the maximum light intensity required for translocation is far less than the corresponding value for photosynthesis. This lead them to the development of the concept of phototranslocation which operates in unison with photosynthesis.

3. Oxygen : Oxygen seems to be a necessary factor for translocation particularly during vein loading as it is beleived to be an active process.

4. Moisture stress : Moisture stress (scarcity) seems to have a retarding effect on translocation. Plant and Reinhold (1964) have found that scarcity of moisture affects solute transport out of the leaf more than the movement within the vasculature.

5. Metabolic effects : Vein loading and unloading are believed to be active processes involving metabolic energy, hence they are intimately associated with the respiratory state of the plant. All the factors that influence the rate of respiration also affect the rate of translocation. It has also been found out that factors which affect growth and hormone functioning may also influence translocation.

Internal control of translocation : Patterns of translocation are not hapazard but precisely controlled with a definite relationship between source (supply) and sink (consumption). Apart from the physical factors of the environment and internal relationship of concentration gradient, the overall metabolism of the plant seems to be having an integrated control on translocation. The movement of solutes in any direction is not

merely on the basis of a concentration gradient. It is possible that hormones of growth may regulate and ediate the nutritional traffic to bring about a coordinated growth. While apparently translocation may appear to be the movement of solutes based on diffusion gradients, the underlying mechanism which generates these gradients maybe an integral part of a complex mechanism involving the whole of plant metabolism. The interrelationship between photosynthesis and translocation is an area of plant physiology that merits serious research.